Circuit-Bending

Circuit-Bending

Build Your Own Alien Instruments

Reed Ghazala

Wiley Publishing, Inc.

Circuit-Bending: Build Your Own Alien Instruments

Published by
Wiley Publishing, Inc.
10475 Crosspoint Boulevard
Indianapolis, IN 46256
www.wiley.com

ISBN-13: 978-0-7645-8887-7
ISBN-10: 0-7645-8887-7

Manufactured in the United States of America

1 0 9 8 7 6 5 4 3

1MA/QU/QY/QV/IN

For general information on our other products and services or to obtain technical support, please contact our Customer Care Department within the U.S. at (800) 762-2974, outside the U.S. at (317) 572-3993 or fax (317) 572-4002.

Wiley also publishes its books in a variety of electronic formats. Some content that appears in print may not be available in electronic books.

Library of Congress Cataloging-in-Publication Data

Ghazala, Reed, 1953-
  Circuit-bending : build your own alien instruments / Reed Ghazala.
    p. cm.
  Includes index.
  ISBN-13: 978-0-7645-8887-7 (paper/website)
  ISBN-10: 0-7645-8887-7 (paper/website)
  1. Electronic musical instruments—Construction. I. Title.
  ML1092.G5 2005
  786.7'192—dc22
                                    2005010349

About the Author

Author and Moonbear outside The Anti-Theory Workshop

Qubais Reed Ghazala (Cincinnati, Ohio, 1953) was first instructed in painting, theatre, and classical piano as a young child (including preschool art courses at the nearby art academy). By the time Ghazala reached the age of twelve, his interest had turned to experimental art (mazes, luminous painting, mirror assemblage, freezing liquids within expandable envelopes—winning awards in local exhibitions for his cubist sculptures). Ghazala soon abandoned formal schooling in favor of his interests, becoming self-taught in his fields.

Besides circuit-bending, Ghazala also pioneers work in photography (lens-less imaging with dye-migration materials; time exposure with cross-axis polarization and aperture shift), magnetic assemblage (kinetic timepieces, anti-gravity sculptures), and aeronautics (an operative flying saucer; various kites, aircraft, and solid fuel rockets). Ghazala's further experimental arts include electronic smoke and liquid/gel chambers, lenticular art, plasma chambers, laser and neon chambers, film and computer animation, video documentary, 3-D drawing and photography, program bending (chance data entry), rubber stamps (as a sampling medium to alter), and woodland art for only forest animals to observe (natural mobiles and interlocking keystone sculptures).

Ghazala is the author of numerous articles on electronics and experimental musical instruments (including the seminal 20-article circuit-bending series in *Experimental Musical Instruments* magazine, 1992–98). Ghazala is one of the Web's most influential cyber-professors, also teaching in person via public workshops and one-on-one apprenticeship programs (circuit-bending and Japanese suminagashi, an ancient chance art).

Although Ghazala is primarily a music composer and instrument designer, he continues to work as sculptor, photographer, visual artist, and author, and in his spare time pursues hobbies of theory-true designing/hacking, building toys, robots, and things that fly; designing mobiles and kinetic electro-lumia; working in optical lens construction; practicing kitchen and deep woods cooking; designing/building home additions; canoeing; hunting mushrooms; studying entomology, mineralogy, and meteorology; star gazing; compiling a home library; exploring natural incenses and resins, psychedelic and experimental music, unusual lighting, rare musical instruments, antique postcards, home-designed postcards, strange medical devices, parlor and stage magic, ESP, unusual natural phenomena, and the collecting of unidentifiable objects.

Ghazala's work has been covered in all the world's major press (including, in the United States, the *New York Times*, the *Wall Street Journal*, the *Washington Post*, *The Smithsonian*, and *WIRED*); his work is kept in the permanent collections of The Museum of Modern Art, the Guggenheim, and the Whitney in New York City; and his instruments have been designed for some of the planet's best-known and most influential musicians (including Peter Gabriel, Tom Waits, the Rolling Stones, King Crimson, Faust, Blur, and many others).

Ghazala is recognized as the father of circuit-bending and is credited with launching the first grassroots electronic art movement.

Credits

Executive Editor
Chris Webb

Development Editor
Ed Connor

Copy Editor
Susan Christophersen

Production Manager
Tim Tate

Editorial Manager
Mary Beth Wakefield

Vice President and Executive Group Publisher
Richard Swadley

Vice President and Publisher
Joseph B. Wikert

Project Coordinator
Erin Smith

Graphics and Production Specialists
Karl Brandt
Jonelle Burns
Kelly Emkow
Denny Hager
Stephanie D. Jumper
Clint Lahnen
Lynsey Osborn
Melanee Prendergast
Brent Savage
Amanda Spagnuolo

Quality Control Technicians
Leeann Harney
Jessica Kramer
Carl William Pierce
Dwight Ramsey

Proofreading and Indexing
TECHBOOKS Production Services

Cover Design
Anthony Bunyan

*To artists working in new fields everywhere
and struggling against all odds.*

Acknowledgments

Who supports the arts?

The big names come to mind, the foundations and institutions whose grants go a long way to endow board-approved art—art whose "viability," often, is measured by an unshy degree of public acceptance. But what about art that is supported more by the maverick mind than the common nail?

History teaches us a few lessons here. Claude Monet's eventual salons made his work known, but impressionism owes immeasurably to friends and fellow painters who sheltered Monet through his many destitute periods. Renoir, the same. And it's well known that Vincent Van Gogh's brother, Theo, kept Vincent happening during the very worst of times. Cézanne, considered by many the best example of the postimpressionist school, found in the simple support of fellow artist Pissarro a reevaluation of his own treatment of color and subject, prompting wide discussion that thereafter became key to interpreting a painting's "space."

Established in this way, via the "grassroots" of family and community, waves of new art were ushered in—cubism, surrealism, expressionism, kinetic and pop art, "abstract" and multimedia art of all types. For the most part, we know of these arts and their artists not because of their later work, as it might seem, but rather because of their early work, unrecognized by the public yet supported by friends and family. And this, not surprisingly, is exactly how it went with circuit-bending.

I've never been offered a grant to circuit-bend or teach, nor have I ever applied for one. In fact, what has now become the worldwide circuit-bending movement was funded only with the spare change in my pocket (the other pocket had a hole in it). Luckily, the technology of circuit-bending required nothing more!

Looking back to the initial days of my discovery process, I find but one aspect of these times that coalesced bending into being: the 1960s counterculture. Were it not for the hippie ideal of "mind expansion," there simply would have been no appreciative audience for my early work. In fact, the Midwest was teeming with individuals not only unaware of experimental art but downright hostile toward anything that challenged the status quo: With regard to stand-still Americanism, "love it or leave it" was the only guiding light offered.

So let me acknowledge that the underground society of these new American patriots, with flowers in their hair and songs of revolution on their tongues, found circuit-bending to be "mind bending," and gave my art the only refuge possible.

My musician and techie friends of the era helped me at the workbench and on stage; they sheltered me and my instruments from unfriendly audiences, and provided a kinship that still lives to this day.

Early band members, such as Mark Shaw and Alan Miller, joyfully accepted my instruments into the performance lineup. Neil Friedman, in his unique wisdom, provided my first circuit with a name. Joshua Litz, friend and neighbor, was a bending partner of mine from the very start—the two of us exploring circuit-bending and together registering the same amazement as we watched the first circuits, back in the 1960s, reveal themselves. Marc Sloan, also a bender and buddy, recorded and soldered with me during the later, but still early, times.

Over the ensuing years, no one has supported me, my ideas, and my work more than my wife, Pamela. More than once she's lived with my one-hour workshop promise turning into an all-afternoon event. She's survived piles of dismantled toys cascading down upon her from workshop shelves. She's put up with endless side trips to dusty warehouses, freezing in the aisles with me in wintertime, sweltering in the summertime, and absorbing more radiation from military surplus vapor lamps than she'd be exposed to on the surface of the sun. She's learned how to search for vintage electronics, glass eyes, specific potentiometers, and square, tri-color, four-lead LEDs. When nothing else seemed to be there for me, Pam was there. I love you, Pam! Without you I would not have made it here, or anywhere at all.

It was a bold move on behalf of Bart Hopkin to include electronic work in a journal initially dedicated to acoustic instruments. But it was here in Bart's publication, *Experimental Musical Instruments* magazine, that circuit-bending first reached the larger audience. I thank Bart for taking this leap, back in the early 1990s, and having faith enough in my art to provide its first ink. My early *EMI* articles also benefited from the assistance of Tony Graff, who typed the writings into text files, giving Bart a much easier day.

Soon after, the Web became the exposure medium of choice for divergent art. Duane Shaw provided me with my initial page online—a simple little window displaying a few bent instruments. But it was the first bent instrument site to be seen in cyberspace anywhere.

Barry Gott (Doc Salvage), of EIO.com, donated the next Web space, and this is where the larger, present site of anti-theory.com was launched. Barry is "no longer with us," though I suspect he'd argue that statement in a moment. It was Barry's hosting that provided the public with the Web's first circuit-bending instructions.

Today, anti-theory.com is hosted by Greg Bossert, of suddensound.com fame. Greg's donation of this space has helped the bending community continue to grow and prosper, and I wish to thank him deeply for his continuing support.

But who's responsible for the creation of anti-theory.com itself, the Web site that launched the entire bending revolution? Although I supplied most pics and animations, it is the tireless work of Nebulagirl, a.k.a. Dr. Cynthia Striley, that stitched the site together and made it what it is today. If I ask Cindy what she thinks about a certain graphic presentation, her answer magically appears online as an example for me to see, and with ten times more input than I ever requested or dreamed of. Cindy, thank you!

Anti-theory.com is, in the end, the result of many people's ideas. Mark Milano, Tommy Dog, everyone on my "Bored of Directors" (aren't we all bored of directors?), and the bending community in general have all played important roles in shaping the site.

Of course, the reason I have this opportunity to thank anyone at all comes back to Chris Webb, my editor at Wiley. Chris and I spoke numerous times prior to the official launching of this book. I swiftly became aware of the pioneering spirit in Chris's mission: to publish texts meant to explore innovative technology and quickly empower readers to take part (this is the successful mission of the entire ExtremeTech series). He and I understood each other immediately, and I can't thank Chris enough for his faith in my work. As did Bart Hopkin, Chris has breached the expected theorems of electronic design and provided readers with a window swung wide open upon an exciting new way to think about and create electronics. Many publishers, strapped to traditional concept, would consider this a frightening prospect.

Working side by side with me throughout this book was my development editor, Ed Connor. Ed, how did you put up with me? Your wise editorial eye saved my writing more than once, you've taken in stride misspellings that would have sent others to the cooler, and discussed things with me as though I were a normal person. Your stoic fortitude, kind demeanor, and inarguable skills will remain a fond memory forever.

Bending the circuits within the projects chapters has been made easy chiefly because of the fine circuit renderings and other illustrations provided by Karl Brandt, Joni Burns, Kelly Emkow, Clint Lahnen, Brent Savage, and Lynsey Osborn.

James Reidel, in both his technical advice and superb poetry, has given me fine inspiration (as well as a fine literary inferiority complex). These lessons are greatly appreciated and I'll stand in line for more.

In my research on the nitty-gritty of this and that of electronics, I've found the simplicity of the Forest Mims books (now out of print, unfortunately) to be very helpful. I've run electronic ideas by so-involved friends, Nebulagirl and Giblet, always at hand with informative feedback.

Without the people who've been drawn into in the storm of bending, the circuit-bending community itself, there would be no reason to write this book. To all of you, I extend a heartfelt thanks. My love of experimental music and instruments was behind my going public with my private workshop notes. I thought that if my teachings could expand the planet's experimental instrumentarium, my work might be worthwhile. You, fellow benders, made this hope a reality. Thank you for adding your own original inventions to this wonderful new field. You've provided the driving force that took bending out of the dark and set it before the (admittedly confused) public.

There are lessons in life so cosmic that they cannot come from a species as tied to the permissions of science as we. For these lessons, some of the most precious I've ever had the privilege to learn, I'd like to eternally thank Moonbear, Merlin, Hukazuwailia, and Pook. Please, keep teaching me.

Finally, I should probably thank my mother for installing those little safety caps in all the unused wall outlets when I was first crawling around and "discovering electricity" with her hairpins.

Introduction

There is that scene in *The Wizard of Oz* where Dorothy's black-and-white world is quickly swept away by a wild tornado, her farmhouse dropped back to earth far, far away from the pig-pens of dusty Kansas. You remember—she then walks to the door of her familiar monotone home and slowly opens it upon a Technicolor world, literally as well as metaphorically leaving the black-and-white of her old Kansas behind.

That's exactly what this book is about. It's *your* doorway to a strange new world, like Dorothy's Oz, filled with fantastic new scenes and characters. It's your way out of black and white.

This book is a first in three important ways. For the Wiley ExtremeTech series, it's the first book to forgo the usual, theory-bound design process in favor of chance electronics. This is wonderful! After all, now everyone's invited to the party.

For me, as writer, it's a first in the sense that I'm abandoning a long-held standard in my teaching. My aim, more than a decade ago when I began to write about the DIY of circuit-bending, was to launch new, *unique* instruments by means of explaining only the general discovery process of circuit-bending instead of using the more standard "this wire goes here" dialogue—a dialogue that usually results in exact duplications of a target instrument.

However, I soon realized that the publishing of my actual schematics, in addition to the general "keep looking" instructions of circuit-bending, produced both variations upon a theme as well as much inspiration to forge forward and design purely self-discovered instruments. Unique instruments.

Reflecting this lesson, in this book I've included both the plans for building some of the most interesting bent instruments in the world as well as a guide to the art itself, a guide that unlocks endless new designs for you to discover on your own, with ease.

Last and most important, this book is a first in that in it I've outlined a new music system, a new discipline, for people to consider. In fact, it's a new voice system in electronic music that, at least in construction, is, ironically, as old as the hills!

But paradox is a standard of circuit-bending. You'll be breaking golden rules of electronic design, theory, and music creation as you explore. And in contrast to Dorothy, you're not dreaming. And you *won't* want to go back home.

What's Wrong with Real Electronic Theory?

Nothing. Understand, I have no ax to grind for true-theory. I use theory when I need it, even if I don't understand it. I'm not in bad company here. Lots of people who play around with electricity don't understand it. In fact, the entire electronics community thought, for decades, that electricity flows from here to there in a circuit. I remembered during that period which way it went, too.

Now they discover that they were all wrong. It flows the other way. And now I don't remember which way it goes at all, thanks to the confusion. What's *really* strange about it is that, even with the theoretical presumptions having been turned upside-down, you can't make ice in a waffle iron or waffles in the freezer. All the things they designed while they were wrong— heaters, coolers, amplifiers, motors, radios—still work just fine somehow.

Simmer down. Whether Ben Franklin was "right" or not, whether electrons are or aren't the only motile element in electrical flow and regardless of this flow's "direction," for most practical applications (and understanding of circuitry) the simple and prevailing models of electrical activity will suffice: Electrons flow from a higher voltage to a lower voltage (or "potential"), or from "negative" to "positive" as in the, uh, *new* conventional wisdom (though "positive" charges—thank you, Ben—still travel the opposite way: positive to negative). Confusing? Well, have no doubt—electricity's still giggling at us all. Cool thing is, as a bender you can laugh right back.

Traditionalists, take heart. Just as bending led me into "real" electronics, many benders report to me the same. As mentioned, bending stirs great interest in electronics, and new designers often follow their curiosity into schooling not otherwise planned. Viva el electron! Or, are electrons female? Now there's a use for the electron microscope I won at sealed-bid auction for one dollar and one cent, surplus, and way too heavy to budge.

Anyway, you'll see true-theory in my Photon Clarinets, my human-voice synthesizers, my insectaphones, and lots of other things I design entirely on my own, "from scratch," looking to IC pin charts, schematics, and the occasional math or table. Doing it the "right" way has helped me create many fine instruments. But I've found that theory is not always the way to think, and *clearly* not the only way to create.

Whom Is This Book For?

In general, this book is for people who want to create a new music by means of designing their own experimental instruments. But it's also for persons who want to understand the new music of circuit-bending from a technically anti-theoretical yet procedural perspective, so as to understand the movement's headspace as well as its design history over many years.

Specifically, this book is for the person who wants to design experimental electronic musical instruments but comes to the bench with no knowledge at all of electronics. So we're talking all technique and just about no theory. This is how bending works and why some traditional designers get mad at me, as though I'm giving away secrets, or making things too easy. In reality I'm just bringing more minds into the fold.

This book is for educators who want to leap a few hurdles and bring into the classroom a fantastically fun introduction to either electronics or experimental music. Bending creates an atmosphere of immediate interest and learning. It provokes a curiosity that often leads students into the wider world of more traditional, theory-true electronics.

Too, this book is for the person who would like to become a well-paid electronic designer overnight. Literally, that is, *overnight*. That's just how fast and easy the techniques are.

The instruments born of circuit-bending are so alien, so unusual in their musical output, that they market to an increasing audience. Prices allow an enterprising individual to earn a decent living. In fact, wages can compete with "real world" jobs, and you don't even have to get dressed.

What Does the Book Cover?

I look at the most interesting aspects of why circuit-bending is important: the "immediate canvas" it presents to explorers, the animal-instrument "BEAsape" creature you'll become when using a body-contact instrument, how you fit into the scheme of things as explorer of a new art, the concepts involved in dealing with living/dying instruments, and the *alien* in alien music engines.

I cover all the exploration techniques involved in finding the strange musical worlds hidden in ordinary circuitry all around us. I cover making instruments light sensitive and touch responsive. I get into implementing line outputs for instrument amplification and recording, thereby exposing a frequency response way beyond what you might expect. I explore adding lights to instruments, designing instruments into unusual cases, adding unique controls, and way, way more.

How Is This Book Organized?

Because no one can predict the outcome of circuit-bending, you're taking a journey into unknown territory here. All you can do is approach with the right attitude and tools to get the best shot at good results. And that's how this book is organized: attitude, tools, and results.

First, I discuss the headspace, the mindset of bending—the concepts that separate it from other design or modification practices. I try to help you understand why bending does what it does and get a feel for how these principles have resulted in a new instrumentarium, in a brand new category of instruments. I consider the issue of experimental art and music. And I discuss your role as an artist/designer on *the threshold of invention*.

Next you gear up for the work. You get your standard as well as specialty tools together. You look at your options and assemble a low-cost workspace that will keep everything rolling smoothly along.

I list all the electronic parts you need and discuss how they work and what effect they'll have on bent circuits. I talk about how to bend circuits, step by step, in every aspect from mechanical to electrical to final finishing techniques. You'll acquire a knowledge powerful enough to turn literally thousands of available circuits into unique alien music engines—all before you even get to the actual project section of the book.

Last, you'll get immediate results. In the Projects section, you build the actual instruments that started the circuit-bending revolution. These projects demonstrate all the most important aspects of circuit-bent instruments and music. I demonstrate simple projects as well as complex ones. I consider deviations on a theme. I have you skate on circuit-bending's thin ice and meet the ghost in the machine.

"How'd Ya Do That?"

This book, in the end, is nothing more than my response to the single most-often-asked question in my life: "How'd ya do that?" Well, here's how I did that. Even so, this art, like all arts, is a try-at-one's-own-risk art. If you follow this heretofore uncharted art, you must assume responsibility for yourself, just as though you were hiking in a strange new woods on your own. And just as I did when I first ventured out on this journey as a 15-year-old, nearly 40 years ago, entirely in the dark. I proceeded with due caution. I am still cautious.

Neither I nor the publisher can assume any liability for what you might meet in this strange new woods. It is also your responsibility to determine whether the use of this information in the construction of instruments for personal pleasure, or for sale, infringes upon any other party's rights or constitutes a legal violation of any kind. I present this information, my journey through strange times and even stranger science, as a source of entertainment as well as education. My hope, simply, is that pleasurable enlightenment will result as I not only answer *how I did that* but, more important, *why I did that*, and how all the pieces of this unlikely story, remarkably, fit together.

Contents at a Glance

Contents

Chapter 7: The Well-Behaved Workspace 75

Part III: Dr. Frankenstein, May We Proceed? 83

Chapter 8: Soldering Your Way to Nirvana 85

Part IV: Eighteen Projects for Creating Your Own Alien Orchestra   197

Chapter 14: Project 1: Original Pushbutton Speak & Spell Incantor . . . 207

Chapter 15: Project 2: Common Speak & Spell Incantor 217

Chapter 16: Project 3: Speak & Read Incantor 225

Exploring Circuit-Bending Today

L ooking back over the past 38 years, and having been circuit-bending throughout, I can tell you one thing with certainty: There is no better time than today to circuit-bend. When I began, there were only a few things easily available to bend, and no instructions were available anywhere. Today, bendable circuits are everywhere—they're dirt cheap in the charity stores, circuit-bending newsgroups are springing up all over, festival organizers are looking for benders, bent instruments are being used by chart-topping musicians, and just look at what you're holding in your hands! Times have sure changed!

Bending the Headspace of Electronic Design

My shop teacher in junior high school was a straight-shootin', all-business, dyed-in-the-wool, only-one-way-to-do-it electronics and woodworking professor with whom I'd argue issues in a heartbeat. He was strictly over-the-counter, and I, strictly counter—"counter-culture," that is. He was quick and sharp; it was rare to leave him speechless.

On the day I brought my first circuit-bent instrument to school (a shorted-out amp now built into a wooden box replete with dials, switches, sensors, patch bay, nuts glued on the lid, and more), my shop teacher was already grouchy and looking for an argument. He approached as I had the instrument out on the workbench. All my classmates had gathered around to listen. I was synthesizing birds, helicopters, and police sirens on the instrument, and running electricity through several people at a time so that we could play the device by touching each others' bare flesh.

My teacher assumed that I knew things I'd not let on to in class (we were making a table lamp out of a bowling pin at the moment, and nowhere in the entire school could you learn synthesizer design). He looked at the dials, looked at the switches, looked at the nuts on the lid, leaned forward, looked me in the eye, and said, "Mr. Ghazala, I didn't know you knew anything about electronics." I leaned forward, looked him straight in the eye, and said, "I don't."

That's the beauty of circuit-bending; anyone can do it. You don't need to be an electronics guru or a shop genius. All you need is the ability to solder and to think outside the box.

Why Bend?

In this day of high-tech electronic synthesis and sampling, why are so many people raiding the second-hand shops, buying and bending yesterday's toys? Good question.

Simplicity: Anyone Can Bend

My shop teacher *was* speechless when I told him, quite truthfully, that I didn't know anything about electronics. I really didn't. I knew a little about soldering, but that was it. That's all I knew then, and with the simple approach this book uses for circuit-bending, that's all you need to know now. And soldering is baby-simple.

Circuit-bending is currently being taught all over the world to people of all ages. MIT has a program teaching grade school kids to bend (imagine—kids learning experimental electronic instrument design at the age I was learning in school to play a plastic flute).

So relax. We won't be getting into scary theory here. You don't need it. As far as electronics goes, you'll be finger painting. Nonetheless, you'll be building some of the most interesting music engines on the planet.

Your Immediate Canvas

At one time, painting was, at least technically, far from the relatively easy task it is today. Today's prestretched canvas and premixed paints, accompanied by brushes ready to buy along with assorted gessoes and sealants, make the modern painter's canvas pretty "immediate." You just walk up to it and paint.

But at one time painters had to craft their own pigments and paints and know much more about the actual science involved in their medium before it was even possible to dip a brush. These nature-gathered raw materials resulted in the palette of the Dutch Masters, a rather subdued range centering on dark, warm siennas and umbras. Makes me wonder what these guys would have done with Day-Glo fluorescent paints. Maybe a bad idea.

Many people fear electronics. If you don't electrocute yourself, then certainly the stack of drab theoretical texts to conquer is daunting all in itself. Circuit-bending changes all this: electronic design is now colorful and nearly instantaneous. No, you're not going to get electrocuted. Yes, we'll use the books—if the workbench needs leveling.

Essentially, to bend a circuit you hold one end of a wire to one circuit point and the other end to another point. That's it! Place the wire upon the circuit in an arbitrary fashion, wherever you want, from here to there on the board. This replicates the pure-chance aspect that launched my first instrument as it shorted out in my desk drawer, and it is still the heart of bending. If you hear an interesting sound, you then solder the wire in place, putting a switch in the center of the wire so that the new sound can be turned on and off. That's pretty immediate!

As with the painter who no longer needs to understand the *science* of pigments to create art, circuit-benders no longer need to understand the *theory* of electronics to design instruments. Finally, with electronics, you can just walk up to it and paint.

The Coconut Concept

At first, this free-for-all we're having with circuitry might seem out of place. Fact is, earthlings *musicalize* things. A coconut washed up on the shore could be struck like the wood block of a percussion set. It could become the shell of a drum, the vessel of a flute, or the resonator of a fiddle. Idiophone, membranophone, aerophone, or chordophone, the simple coconut can be modified to fit all the major instrument groups of the orchestra. Add steel strings and magnetic pickup to the coconut fiddle and you've got the electronic group covered, too (Gibson guitars, give me a call).

Second-hand shops, where I find most of the circuits I bend, are like high-tide lines on a beach. They're high-tide lines for a different ocean—the ocean of western civilization. Instead of coconuts we find here everything else cast overboard by our throw-away society. Circuit-bending sees its circuits as the island native saw the coconut. In fact, in a very real sense, these things are the coconuts of our island. Adapt the coconut, adapt the circuit (see Figure 1-1).

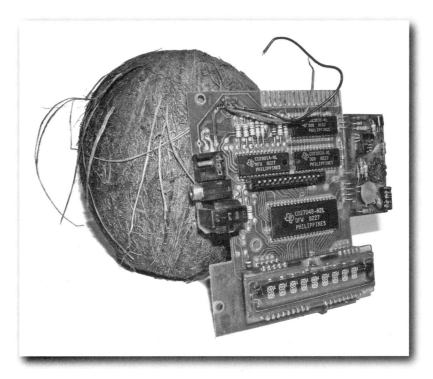

FIGURE 1-1: Circuit and coconut—one and the same

Taking it a little further, and though not a chance art, the chopping and hot-rodding, repainting, rewiring, and renaming of that '57 Chevy you see at the car show is done in the same vein as circuit-bending. Each effort reflects a desire to push the limits of a thing. I recently spoke at a circuit-bending convention in New York City where, as at the car shows, industry dollars propped up the event!

Bending has come a long way from the night in the 1960s when the band and I had to fist-fight our way out of that performance-gone-bad with the Elvis fans at our heels. I recall that we decided not to play there again.

The Threshold of Invention

That the golden age of invention is dead is a romantic notion at best. True, corporate research and development labs seem to outdo the practical capabilities of garage tinkerers. But the truth is that we live in a rich world of discovery, right here, every day. Just depends on where you look.

There is a principle discussed among all inventors, unnamed heretofore. A principle of simultaneous discovery. I discussed it most recently with blacksmith Howard McDaniels of Lebanon, Ohio, a modern smith whose work spans traditional utilitarian repair and fabrication to artistic ironwork to high-precision pieces for the nuclear industry. Howard's example of this principle was how similar developments in metallurgy occurred, unrelated, in various places around the planet at the same time. I call this technological/sociological time-wave the *threshold of invention*.

Although experimental electronic music labs have been around since the 1950s (Columbia-Princeton being the prime example), the music produced then was appreciated mostly within academic circles. This tradition extended fully into the 1960s, relative not only to appreciation of experimental electronic instruments but to experimental design as well.

Important to us is the period of the mid 1960s. Transistorized, battery-powered electronics were now somewhat commonplace, at least in the form of pocket radios and walkie-talkies. Lots of these circuits were susceptible to interference if the circuits were touched, suiting them aptly for the technological side of the equation. What's still missing is the sociological.

Not that experimental electronic music geared for the general public didn't exist prior to the 1960s. It did! You could hear Bebe Barron's self-destructing circuits bubbling away in the movie *Forbidden Planet*, representing the music of the Krell, an extinct master race (something seems wrong there, eh?). Small-town America's Saturday matinees overflowed into the streets with weird electronic oscillations as the assorted sci-fi "B" movies of the 1950s came and went, their otherworldly music colliding head-on with the real-world cold war headlines boxed next to the ticket window.

Little kids were considered as good a target as adults for experimental electronic music in the 1950s. I had a recording that I listened to as a young child, a kid's 45 rpm record, called "Omicron Visits Earth." In this comical fable of a crashed spaceship, Rege Cordic and Co.'s weird electronic oscillator music bridges the story's chapters.

But it was not until the Beach Boys' "Good Vibrations" that a large popular audience had to deal with experimental electronic instrumentation. Musically set as it was, within the Beach Boys record-selling surf-board-rock style, the theremin struck a wavering note with the public, and the electronic instrument stage was set for bigger and better, not to forget louder, things.

Note

Actually, the instrument heard on "Good Vibrations" (and two or three other Beach Boys songs) was not a theremin at all. Instead of being "space controlled" (played without touching) like a real theremin, Bob Whitsell's "Electro-Theremin" was a *mechanically controlled* oscillator. For accuracy's sake, this instrument might be better named the Mechano-Theremin. But then, that wouldn't be a theremin at all, would it? The Electro-Theremin heard on the Beach Boys' hit was hospitalized at the end of its career. It was then the elderly, not beach bunnies, "pickin' up good vibrations" from the instrument: In the hospital wards the Electro-Theremin was used to test the hearing of cranky oldsters, many of whom, no doubt, had heard it in earlier days and promptly turned their radios off.

The 1960s public was ready, as Bob Moog would discover, for electronic synthesizers. Things were happening within corporations and universities—research teams were designing both musical as well as purely experimental electronic instruments (the latter wonderfully exemplified by the Michel Waisvisz/Geert Hamelberg "crackle box" of the late 1960s, an instrument later developed at Steim in Amsterdam that, although from the start designed to do what it does and therefore not a circuit-bent or chance-designed instrument, still uses body contacts for control).

But something very important—more important to us, in fact—was also happening outside the theory-guided teams and institutions. It was happening way down at street level. A totally new approach to design was being launched. Renegade electronics' insurgence was at hand.

The threshold of invention, placing susceptible mid-1960s electronic circuits and music-curious humans upon the stage simultaneously, was at work creating its usual synchronicity of planet-wide discovery. All over the world, lone wolves like me were discovering that the accidental howling one encountered upon touching the live circuits of battery-powered audio electronics was musical.

Where Did Circuit-Bending Start?

Probes are sent into deep space to listen to alien worlds, but alien worlds aren't always that far away. For me, in fact, a portal to an alien world was hidden in my childhood home, right at my fingertips. Here's what happened.

I'd always been interested in unusual sounds and music. I even awoke once actually singing along with an otherworldly choir I was dreaming of, startling my mother awake in the next room. Five-year-olds singing in foreign tongues from somewhere in dreamland? I was delighted, my mother a little less so.

This was the mid-to-late 1950s, the same period in which I was tuning between FM stations to hear the strange squeals, touching the glowing tubes to listen to the weird buzzes, and fingering my walkie-talkie circuits to get that loud and nasty "WAAAA!" sound.

A "Girder and Panel Hydrodynamic Building Set" had me constructing plug-in circuits when I was eight or so, after building electric motors for plastic models for a few years prior. I became enamored enough with electricity to necessitate the safety capping of all unused wall outlets lest I repeat my paper clip stunt, which I was, luckily, caught at while performing its debut.

I spent my best moments during these years whistling through vacuum cleaner hoses and talking through spinning fans, hitting telephone poles with hammers (press your ear to the pole right after—try it!), and all the other noisy stuff kids do in their off-time, all day long. And I listened to the distant world through my "Big Ear," an electrified parabolic sound dish, always pointed toward the future, toward what the 1960s was about to bring.

Synthesizer music, in the mid-to-late 1960s, was all the rage. It was new to most people and very, very cool—especially if you were a 15-year-old hippie-musician, as was I! However, my friends and I were all penniless, and the possibility of our getting hold of a synthesizer was certainly remote.

So we listened to the oddball electronics of Silver Apples, Aorta's "special effects," Joseph Byrd and Gordon Marron's electronics on the seminal *United States of America* album, plus *Switched on Bach's* new Moogs as well as the earlier music of such experimental electronic composers as Bülent Arel, Mario Davidovsky, and Vladimir Ussachevsky working out of the Columbia-Princeton Electronic Music Center. And we wished we had a synth.

My high school desk drawer, circa 1967, was my junkyard. In it I had a wrecked Radio Shack mini amp. This was a nine-volt, battery-powered transistor amp containing no integrated circuits (ICs) at all. Inside were just a few small transformers, transistors, resistors, and capacitors, all gathered around a small central speaker. The gray plastic case was no larger than the palm of your hand and looked as uneventful as a gray, overcast sky.

With the muses of blind luck on my side I'd left the battery in, the back off, and the power on. In pursuit of some other item in the drawer, for a project now lost in time, I'd pulled the drawer open to rummage around. All my cool stuff was in there, nestled around the little gray amplifier that was about to change my world, just as it is about to change yours.

I closed the drawer and was immediately in the midst of some of the most unusual sounds I'd ever heard. Why? By pure accident, some unknown metal object had fallen against the exposed circuitry of the amp, shorting-out the electronics.

Of course, I didn't know this at first as I looked at the stereo system (it was turned off) and then all around the room to try to figure out what could possibly be making the extraordinary noises. A "flanged" pitch was sweeping upward to a higher frequency, over and over again, sounding like a miniature version of the massive Columbia-Princeton synthesizers. But I didn't own a synthesizer! Or did I?

Finally opening the drawer and realizing that fate had created for me a mini synth from my toy transistorized amplifier, two thoughts immediately struck. First, if this can happen by means of an accidental short circuit, what might happen by shorting things on purpose? And second, if this can happen to an amplifier, a circuit not meant to create a sound on its own, what might happen if you shorted circuits designed to create their own sounds already, such as toy keyboards and electronic games? Though I didn't name the design concept of the found-by-chance creative short circuit until 25 years later, "circuit-bending" was born.

I soon discovered many different creative short circuits within the mini amp's circuitry by using a hair pin to span the circuit, point to point. Additionally I found that all kinds of interesting variations in sound would occur when I ran these new circuits through various electronic components, things such as potentiometers, photo cells, and capacitors. A series of instruments resulted as this original circuit was rebuilt into various configurations to allow more room for these electronic modifications.

The first was the most unusual. Designed and redesigned during the period of late 1966 through early 1967, it settled into an aluminum foil body-contact and patch-bay instrument. Yes, with spinning speakers (see Figure 1-2).

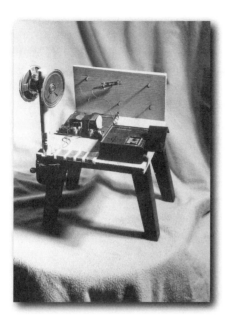

FIGURE 1-2: Ghazala's first circuit-bent instrument,
c. 1967–1968 (mock-up; the original was
destroyed by an irate audience)

I'd found that simply by touching the bare circuit with my fingers, I could get it to squeal, and even—with careful adjustment of finger pressure—actually sweep the pitch with a degree of control. By soldering wires to these body-contact points, I extended them to aluminum foil pads on the instrument's front control surface so that they could be touched with ease.

Then, I implemented all the interesting point-to-point short circuits I'd discovered in a patch-bay fashion using a backboard on the instrument. Nails were driven through the backboard. A wire was then soldered to the extending tip of each nail, behind the backboard, and these wires were then soldered to sensitive circuit points.

Another wire was then soldered to a final point on the circuit, a point that when connected to any of the other points, which were now extended to the nail heads on the patch bay backboard, would result in new sounds. To this lone wire I soldered an alligator clip, enabling the wire to be clipped to any nail head on the patch bay, thereby making the circuit-bending connection and creating the new sound.

The amp's original speaker was replaced by a pair of similar speakers, now attached to opposite sides of a wooden dowel. Mounted onto the axle of a powerful slot car motor, the speakers could be spun using the slot car's original motor and accelerator pedal, mounted now on the instrument's playing surface.

In my neck of the woods, in the late 1960s, a 15-year-old hippie kid playing a prototype synthesizer with tiny spinning speakers was not always embraced with critical enthusiasm. Especially if playing in a neighborhood church to rowdy Elvis fans. The resulting tussle damaged this fragile instrument to the point that all I could salvage was the circuit.

Certainly not recognizing this as a classic incident of experimental music–meets–John Doe, instead thinking practicality, I simply redesigned the circuit into an irate public–resistant case. And so it survives to this day.

The new instrument, finished in 1968, was housed within a lidded cedar box. I'd glued whole nuts, in the shell, inside the lid to accommodate the patch cords that could, in one way or another, be wrapped around the nuts to keep them close at hand while still out of the way. (Well, you work with what ya got, right?) Thanks to a friend who enjoys surrealism as much as boysenberry juice, this instrument became known as "The Odor Box" (see Figure 1-3).

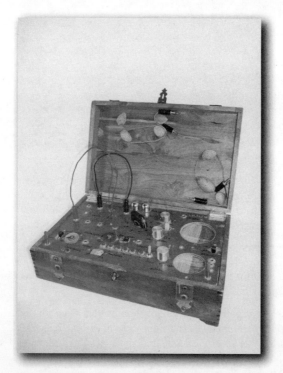

FIGURE **1-3: The Odor Box containing the original bent circuit, now audience-proof, c.1968**

The patch bay was now implemented with ⅛" audio jacks on short cords instead of with the alligator clip and nail system. No, they didn't carry audio, just the shorts I'd discovered. Also new to this circuit was a rotary switch allowing the shorts to be sent through a collection of resistors and potentiometers as well as a photo cell. The aluminum foil body contacts were now replaced with solid metal domes. I added a separate toy organ circuit to experiment with polyphony and beat notes (the pulsing you hear when two frequencies are close to the same).

Also added were a separate amp and line outputs for recording and amplification. Later I added light-emitting diodes (LEDs) as soon as they hit the consumer marketplace.

Yet another version from the same time period was my portable body-contact-only model. Containing eight foil contacts, it could make all kinds of bird and insect noises (see Figure 1-4).

FIGURE 1-4: The Cat Box, a slimmed-down, meowing and chirping, all-body-contact instrument, c. 1968

Because it was simply unheard of back then, in the 1960s, to carry around pocket-size electronic bird synthesizers, holding the instrument under a restaurant counter and making it chirp like a song sparrow produced a fine audience. Especially if you pointed up toward the light fixtures at the same time and said things like "There he is!"

Technology produced more and more complex sound circuits, and my orchestra grew and became diversified. Alone in my suburban basement, I relentlessly experimented in my ad-hoc sound lab. I made *working* science fiction instruments while *Star Trek* debuted on TV, and while *Lost in Space*, *The Jetsons*, *2001: A Space Odyssey*, *Close Encounters of the Third Kind*, *Star Wars*, and so many other sci-fi episodes special-effected instruments into existence. And I listened, often amazed, to what seemed to be a new, alien world of music.

Here in the United States, along with myself, John Hajeski had also fallen down the same 1960s rabbit hole. Admiring the ghost in the machine, John created his "portable anarchy" instrument: a suitcase of circuit-bent radios that, via body-contacts, like my first instruments of the same era, could make experimental electronic music with the touch of a finger to the circuit. John and I weren't aware of our similar experiments. As with us and so many others who certainly heard and recognized the creative short circuit back then, the threshold of invention swept us all along like so many voided warranties in the wind.

Enter EMI

It was in *Experimental Musical Instruments* magazine, in 1992, that I first published the term "circuit-bending" in an attempt to help identify the emerging art, open it to discussion, and differentiate it from the more expected and theory-true design processes. Ever since then, the art has been gaining momentum and demanding clarification while spreading, like alien musical bacteria, throughout the planet's circuitry. The result is at least a design revolution in electronics if not, as the press has responded, "the planet's first electronic art movement," the "first electronic art-object movement," or even the "first grassroots electronic music movement."

Doesn't really matter. What matters is that you're here, as was the EMI audience before the art was afforded any major press at all, and that we're at the beginning of the era of circuit-bending. As did the school of impressionism, bending will branch and produce numerous artists who will work in varied styles. Most of the new music and instruments still await discovery. The field is *exceptionally* wide open.

Think about it. In one easy evening you can discover and build an instrument capable of making sounds and music no one else has ever yet heard, an instrument that exists nowhere else in the universe. Within circuit-bending's threshold of invention, the golden age of discovery is not dead at all! In fact, you're about to learn enough about circuit-bending to do some inventing of your own. The following chapter explores circuit-bending's new instruments and their often unbelievable behavior. You'll be introduced to the concept of *chance* electronics, the mechanics of *clear illogic*, and how you yourself can become a new species by actually melding your body with the electronics of the circuit. You'll discover a new way to look at the lifespan not only of bent instrumemts but also all musical instruments. Most important, the next chapter takes you inside circuit-bending's garden, an electronic zoological garden where you'll meet the creature instruments you're about to design.

Circuit-Bending's New Instrumentarium

My boyhood home was in a quiet middle class neighborhood where, on occasion, something other than the Mister Softee ice cream truck would disturb the peace. Upon such a summer day an insistent knocking at the front door drew me away from my 1968-or-so recording "studio" (repaired Wollensac 7" reel-to-reel mono tape deck that also served as the left channel of my stereo system). I'd been recording my first instrument, the bent mini amp, housed now in the cedar box.

My friend Josh from up the street was at the door wondering whether he could borrow a tape of the instrument for his dad. For his dad? It came as a surprise to me that Mr. Litz was a fan.

Turns out his neighbor's stereo was too loud, and the ensuing wattage war was gaining ground for neither party. Mr. Litz had decided it was not the powder charge but rather the payload that needed consideration. And because the police had not yet arrived . . .

Understand, I thought that the weapon I lent for the battle, a recording to be blasted out of the Litz sound system, was a nice piece of experimental work. It included some fine full-range oscillator sweeps, lots of weird swishing noises, lots of experimental chatter, and, I thought, a sensitive regard to dynamics and flow.

Mr. Litz played my work at full volume for a few minutes, 300 watts per channel, with speakers now hoisted up into his windows and pointed toward the apartment building right across the walkway. He then hit the pause control to see whether the recording had produced the desired result. The outcome of my composition, at full blast, seemed to have had a tranquilizing effect upon Mr. Litz's neighbors. The apartment across the way was silent. I did have a new fan.

Alien Music Engines

Alien music engines. That's a cool title, isn't it? Makes you think about flying saucers and music truly from out of this world. Well, guess what?

In the chief ways that establish music via instrument design, circuit-bending produces a different animal and a *truly alien* music engine. In effect, the bent instrument opens a world of music that in theory, circuit design and, often, musical composition operate outside of human presumption. And presumption in music is key.

When instruments are developed the first presumption is usually upon scale: What notes should it play? In an instrument with set notes, such as a piano, this decision is crucial. In a "normalized" synthesizer we also center design around, in most cases, the equal-tempered scale, those same 12 notes up and down the keyboard.

Next, we presume upon the design process exactly what we need to do to create an instrument, be it acoustic or electronic, which makes complete sense if you know where you want to go. We need the piano hammers to strike when the associated key is pressed, and we need them to hit strings that are tuned to the pitch that their placement on the keyboard *presumes* that they should. For the synth we'll need oscillators that can be filtered to sound like the tones decided in advance that we want, the tones we'll have printed on the case next to the tone buttons: bells, strings, flutes, all the things we *presume* people want to hear.

What makes alien worlds alien is that you can't presume what's there. They operate outside the world of human knowing. You cannot presume what they're going to do. This is why we send probes to alien worlds: We don't know what's there. But we want to.

In a very similar way, circuit-bending probes the circuitry to hear its intrinsic music, allowing it a personal prose beyond programmed recitation. Now the circuit is operating way outside of the original designer's plans, by chance shorted out into a robust new language machine, and human presumption no longer pulls the strings.

We do have an alien music engine here. Until the saucer lands in your back yard and takes you to a new world, let bending take you there.

Chance Electronics

Here is the great leap. The wonderful leap that finally sees electronics as an organic mechanism, animal-like, and listens to it as though through focused sonic X-rays falling, like rain, by chance, here and there upon the circuit.

Although this sounds wild, the concept is not new at all. *Chance art* is a wonderful field with some of the most beautiful examples dating back much further in history than the purpose-bound works with which we are usually taught about art, and by a people more culturally advanced than the Europeans of the same period, whose societies would not produce Leonardos and Monets for centuries still.

Though certainly in existence earlier, the first known examples of suminagashi are dated to Japan during the Heian era (794–1185). This exquisite art, one of floating ink on pure water, produces sublime images frozen in their metamorphosis and captured forever by the printing paper, dipped carefully upon the liquid surface (see Figure 2-1).

FIGURE 2-1: Suminagashi by Ghazala, 2002

This art was held supreme by the Japanese people, the paper so valued that gold flakes were integrated into the material and only the highest poetry was penned upon it, the script following the surreal lines of ink.

Chance art is behind the changing configurations of Calder's mobiles and many other kinetic pieces, and is certainly behind the work of many abstract painters. We don't need to go as far as the action painting of the American abstract expressionists for an example. Just look how oil paints might blend within a single impressionist stroke of even a controlled painting, or how watercolors bleed out into the paper and into each other, quite freely, to define their medium. To one degree or another, art is often chancy.

We needn't stop there. I'll ask you: Is the Art Nouveau style a chance art? No? Well, I suppose not. But the designs and motifs of Art Nouveau, and to a lesser degree Art Deco, were inspired by the lines of nature, master of chance art. When we "naturalize" a planting by throwing seeds upon a hillside instead of planting in neat rows, we're trying to replicate nature's chance arrangements. It is these chance arrangements in nature, the rhythms of shape and color surrounding an early and uninsulated humankind, that we suppose to be behind all our art.

Chance is a powerful, creative force whose "accidents" have not only provided us with art but also the discovery of penicillin, infrared and, well, Velcro. Truth is, chance discovery is the root of much of our understanding of the natural laws that govern our very existence—despite the scientists who'd have you believe they're in control of it all. Sissies.

Circuit-bending taps this powerful, creative, wildly artistic force, finally bringing chance to the realm of electronic design and *you*. As discovered, so it proceeds within the unusual technology of circuit-bending.

Fuzzy Logic vs. Clear Illogic:
The Law of Opposites and Bent Science

That's a mouthful! As you've heard by now (or learned by the ice cream headache), for every thing there is an opposite thing. Sometimes you use a thing to guess what its opposite thing is, especially if you can see only one of the things. The same, I'm afraid, goes for systems.

Fuzzy logic is a system, implemented by tumbling equations, that is meant to seek a norm within chaos. The law of opposites suggests that a system, a dark bizarro world system, should exist—a system that operates in reverse. There should be a system that seeks chaos within the norm. If fuzzy logic exists, *clear illogic* must also.

We'd also think that, if we're still on the trail of opposites, the "tumbling equations" of fuzzy logic might interpret into nonreason within clear illogic. And, of course, it does. Like it or not, this is how science works, and science always has the last laugh.

What we've discovered is that circuit-bending is an act of clear illogic, itself a scientific *system of application* following the natural laws of chance. We are modeling natural creative forces when we chance-bend a circuit. In a sense, we play God here, so powerful are the creative forces we're dealing with. We provide that mysterious catalyst when we bend a circuit: that spark that changes things, that lightning strike that just might configure life.

BEAsapes

You already thought I'd gone too far, right? Anti-theory, clear illogic, and playing God with circuit boards. How about if I told you that I've had to name a new species in order to discuss one of circuit-bending's most interesting aspects? Such is the case. See what you think.

Early on, way back in the mid-1960s, I began to suspect a new, living, actual flesh-and-blood creature in my midst. And I was part of it. As I made sounds by touching the body contacts on the shorted-out mini amp, the electricity of both bodies, myself and the amp, became intertwined. I was, in effect, a 15-year-old Midwestern hippie proto-Borg.

With both hands on the amp's circuitry, electricity was flowing out of one side of the circuit, through me, and back into the other side of the circuit. In this configuration I was as much a part of the circuit as any component soldered in place. I couldn't see where either the amp or I began or ended. We *were* one.

This new creature needed a name that flowed better than "15-year-old Midwestern hippie proto-Borg." I call this animal a BEAsape. "BEA" is pronounced as in "be a" sport; "sape" rhymes with "grape." BEAsape stands for Bio-Electronic Audiosapian. Instrument or animal, hybrid or mutant, musically as well as zoologically we clearly have a horse of a different color. Yes, the BEAsape's material is temporary, its existence momentary. Like you and me.

Living Instruments

We needn't go as far as the BEAsape to find what we might term "living instruments" within circuit-bending. There are times when bending causes safe thermal limits in circuitry to be exceeded. A new connection might cause too much voltage to pass through a component somewhere, slowly burning it out. Again like you and I, such an instrument will have a voice evolution over its life, from birth to death, as it electronically consumes its accelerated timeline.

We can accept this in ourselves, our families, and our friends. But can we accept this life cycle in our instruments? Until now, we expected an instrument to do tomorrow what it does today or it goes to the repair shop. If the voice changes, well, something's wrong. Right? But what if we *did* apply this principle to the other living things in our worlds. Friends. Family. Pets. House plants. Age, and you're outta here! That's pretty ridiculous.

So we are faced with another new beast! Inorganic this time, as opposed to the BEAsape. Yet by all practical purposes a living instrument. Complete with life cycle. Don't turn it on and extend its life? Spend its life to let it sing? What would you do?

Again, we are faced with what seems to be a unique problem. But it is actually a very common problem of instruments. The real issue here is more a matter of scale.

Classic Instruments Are Not Dead!

That's right. The common instruments of the orchestra are "living" instruments, too. As they age, their voices change, but not, of course, on the time scale of living, circuit-bent instruments. In the world of acoustic instruments this is both good and bad.

The upside of this predicament results in instruments that sound better over time as they age. Woodwind instruments claim this fortune. More famous, though, is the Stradivarius violin, whose mysterious lacquer, it is supposed, has strengthened and purified the instrument's tone over time.

The downside of instrument aging extends beyond undesirable voice change. Although an aging acoustic instrument's voice certainly does change as its joints loosen and alignments drift, a walk through any antique instrument museum will inform you as to which instrument's voices changed the most, and without even hearing them. In fact, you can't hear them. You can't even see them. The instruments whose voices changed the most are the ones that aren't there.

A perfect example of this is the antique harp. There are records, of course, and even pictures of harps of antiquity. But although other instruments of the same era survived, surviving harps are rare. Why? Because they, like living bent instruments, self-destructed because of forces self-exerted upon their structure. The tension of the strings was simply too much for the harp's wooden frame to bear.

The living circuit-bent instrument, then, is not alone in history. And although bad for the harp, perhaps in an experimental instrument whose performance has not been mandated by presumption, the aspect of an evolving voice might even translate in the direction of the Stradivarius violin.

Unique Instruments and Sounds

We wondered a little while back: Why bend? As easy as bending is, as conceptually fascinating as living instruments and BEAsapes might be, and significant as bending might be on the larger scales of invention and electronic music history, people bend because of the unorthodox sounds. These strange sounds alone have kept me at the bench for 38 years, nonstop.

I should admit that my house is a confused museum of instruments, hundreds and hundreds, all centering on the unusual. Mellotron to didgredoo, Stylophone to erhu—they're all here. I design acoustic instruments as well as electronic instruments and am pretty familiar with the voice systems of both camps. In electronics I've built theory-true instruments to explore chance music using deep pseudo-random algorithms, and I've designed many synthesizers to create usual as well as unusual animal languages.

My favorite acoustic experiment was done on my upright piano, which now has an extra pedal. When pressed, the pedal applies a stop to the middle of the bass course of strings, bringing out the second harmonics. You've seen this done on a guitar: A finger is lightly laid across the strings at the various harmonic points while the strings are strummed. On the piano this changes the usual string sound into a bell-like tone, and the piano begins to sound like a carillon. Varying pressure on the pedal yields other tonal effects (see Figure 2-2).

FIGURE 2-2: Harmonic capo on upright piano

It is true—my house is so stuffed with instruments that every catastrophe sounds musical. But because of this close association with worldwide instruments, experimental and non, I can tell you something with complete certainty: Circuit-bending takes you to a new place.

Persons excited by the history of musical instruments appreciate that the clear illogic of chance electronics is somewhat groundbreaking. And the extra-musical implications of invention and sociology are interesting, too. But what's truly fascinating is that anyone at all now can, in an easy evening's work, discover and build a unique instrument capable of making sounds and music outside systems of human presumption, music that no one else has yet heard or even dreamed of.

Considering a New Music

The 1960s was a "clothing optional" decade. What to do with all that bare flesh? Perfume it, sunbathe it, paint it, electrify it!

And that's what we did. You see, the bent mini amp's body contacts, when bridged by a person's fingers, would make the instrument respond as an electronic oscillator. Varying the pressure of finger contact would tune the oscillator in the same way that turning the pitch dial on an oscillator would, from super low to ultra high.

But what was even more interesting was that the electricity could be passed through many people at one time. Clad or semi-clad, we'd have people hold hands, persons at each end of the human chain touching a body-contact, thereby placing everyone within the live circuit. The bent mini amp would respond with a loud, sustained pitch as long as nobody moved.

Whichever persons broke the chain could then play each others' bodies, because various body parts sounded different depending on the complete-ness of contact. A bare arm, for example, made a good contact and resulted in a pure tone, whereas a hairy arm made poorer contact and resulted in a scratchy tone. Moisture made for nice contacts. Kissing even better. So there was a lot of hugging, smooching, slurping, and worse, all to the tune of an oscillator gone wild as we gyrated around the floor, holding onto each others' whatever.

Sure, this was art. But was it music?

Dealing with a New Tonal Palette

Although many music writers quickly become mired in the complexities of what music is or should be, I don't. This has always been an easy subject for me because I never look to category over emotion. As with every color on the painter's palette, every sound elicits an emotion. This in itself transcends style.

All music is sculpture. Whether using the equal tempered scale and 4/4 measure or free-form music "concrete," it's sculpture. Although most popu-lar music is often dreadfully beat oriented, the very nature of sound asks to be seen also as free poetry, open prose. It's not just sound constructed to dance to, to sell to the masses, but also an emotional construct, devoid of economical pressure or "dance-ability," existing instead as storyteller, perhaps, guiding us by sound much as a painter guides us by means of image.

Do we dance to paintings? Must we to music? Are these phenomena so different? If we don't dance to music, must it still be orchestral arrangements, operas, hymns, and other such "programs"? Understanding music and sound in this way, freeing it from "popular" structures, allows its fabric to further unfurl.

One of my greatest lessons here occurred in study hall back in high school, though it had little to do with the curricula. I entered into the quiet of the immense auditorium, found a seat, and settled down to fold the propaganda that my civics teacher had handed out into paper airplanes that could reach the stage, 40 seats away.

Suddenly the air conditioning system shut off and the room actually did become quiet. I noticed an immediate reaction to the cessation of sound. Not realizing that I'd been tense at all, I clearly became relaxed when the sound stopped. All sounds, we know now, affect our emotions whether we are conscious of the sounds individually or not.

To me, this prompts much thought about the power of composing music, a wave-form phenomenon, as one might use color in painting. Another wave-form phenomena, color arrangement, is accepted, even recognized, in paintings from traditional to abstract. Extending this idea to sound encourages one to seek the pure emotional value of sonic tone as painted hue. All compositional limitations are removed; all musical combinations are invited. Alien music, at least alien to earthling conventions, can be *heard*.

Timbre and Chance

In circuit-bending, we'll find new sounds in two main areas: tonal variations to a pitch, and assorted aleatoric, or chance, composition. Both of these can be truly fascinating as generated by the out-of-theory electronic instrument.

The Power of Tone

The first category, tone, is familiar to everyone already through the tone controls on your stereo. Bass and treble. But fuzz is also a tone variant, and similar distortions abound within bendable circuits. You'll also find interesting phasing and flanging effects added to the basic sounds of the circuit as you try different bends.

Although some of these modulations can be very strange, we're still in familiar enough territory, and Jimi Hendrix's guitar in *Purple Haze* will still sound as though it's hitting the right notes no matter the tone. Essentially, when talking tone we're talking timbre, the reason a violin and circular saw sound different when playing the same note. It is the timbre, or wave-form, difference in these two voices that effects us emotionally one way or the other. So let the bent composer be able and the audience be warned!

Experimental and Chance Music

It is the second category, that of aleatoric (chance) music, in which things get a bit more difficult. But we must remember that every series of sounds is a poem, musical or not. And again, every sound elicits an emotion.

Although bent instruments are being used all the time in beat music (some are perfectly suited for this), their "outside" nature asks to be considered within a purely experimental framework. And in many bent instruments, you'll discover some of the most perplexing yet productive chance music engines available.

Fact is, aleatoric music isn't so strange. We hear sound poems all day long. Wind chimes and birdsong are aleatoric music. The same is true with the sounds of wind and rain. Enter train horn and uneven track rhythm, bicycle bell and then factory whistle in the distance, far, far, away. This is a natural, environmental chance composition, a countryside sound poem.

Now imagine that, in place of the players just mentioned, a human hand had the chance to substitute passages of aleatoric music self-composed on circuit-bent instruments into the mix. Instead of the train, we hear an abstract musical language appear and fade away; the train horn now becomes an impossible animal cry from a bent barnyard toy. The background wind and rain are now washes of mildly stuttering syncopated static, and so on down the line.

I've enjoyed pondering the emotional value of instruments. Why are certain combos so effective? The chamber orchestra or string quartet, for example? The jazz trio or the bass-drums-rhythm-lead, four-person rock group? If there's a well-working "chord" of emotional value struck here within these instrument sets, why? The next question is, again as in the countryside example: Can we substitute? What sounds can take the place of the bass/drums/guitars in the rock group to attempt to achieve the same emotional value?

Experimental electronic music has evolved way beyond the monumental oscillator sweeps of the 1950s, entering new spaces today at a startling pace. No longer confined to academia, experimentalism has taken flight and can be heard within many popular genres. Circuit-benders are at the very forefront of this experience of new experimentalism, constantly pushing music forward with original discoveries.

Bending for Business

I collect instruments everywhere for bending: second-hand shops, flea markets, going-out-of-business tents, the classifieds, and yard sales. Recently, at a rural yard sale I happened upon a Speak & Spell for sale. The sticker on it said 50 cents.

Handing over two quarters to the elderly gent at the folding table, I was asked just what I was going to do with the toy, already appearing, at least to him, I suppose, that I was capable of spelling. Well, let him think what he wants!

I answered that I was going to resell it for "two-fifty." Nodding approval, he replied, "Two dollars ain't a bad profit." As I said, let him think what he wants.

What Makes a Good Sales Instrument?

Although circuit-bending has become the mother of all cottage industries, some focus is still needed to do good shopkeeping and have the right instruments in stock. Outside of fair business mission and the usual book-keeping and advertisements (check out any small business guide for these items), some bending-centric considerations are in order.

Availability

If you're making your own instrument from scratch, parts worries are not the concern that they are when bending no-longer-produced circuits (as in the Speak & Spell line). My targets are usually secondhand, often inter-vening between circuit and landfill. In this field there are items that show up regularly enough, and in fine condition, to be standards in my sales gal-leries. But if you're specializing in an instrument or two, finding a good stock of target circuits can be difficult.

You might get lucky and find a store (or Web site) clearing-out a load of "dead" stock. I've bought Speak & Reads, Touch & Tells, Casio SA2s, and other target items this way, a shelf at a time. Find the manager and make an offer. Be ready to be flexible.

As for the rest of your stock of general parts, be sure that any special parts, including cases, are available in quantity, and stock up while you can. As far as run-of-the-mill parts go (switches, pots, and so on), again, find a good-quality part and try to buy in quantity. For example, the last time I bought both pushbutton and toggle switches, I did so by weight, by the thousand, from a surplus warehouse. Doing so brought the price down to 35 cents each.

Specializing

Specializing, as in all business, can be a very good thing after you become known as "the man." You will, however, need an ongoing source for instrument parts. Internet searches can be a big help here. Asking store managers to sell you all the remaining something-or-others is a good move, too, as mentioned. People also buy new. I personally avoid buying new toys because I have a bad feeling about the current state of the production side of the industry.

Sounds

Whether you're specializing on a single design or selling an assortment of instruments, just as in regular instruments, sound is of utmost importance. Cool looks alone won't cut it. Whether tonal variations or chance music are produced, thinking "hi-fi" won't hurt your business plan.

Pricing

I base my prices on how long the job takes. Some people ask for show-stopping museum pieces—one-offs or limited editions. These can take weeks to finish. People also ask for nicely painted series instruments as well as completely unpainted, "bare bend" basics.

For my deepest designs I had a renowned art gallery, one representing top electronic designers (such as Nam Jun Paik) come and appraise my work. The painted series instruments (Incantors, Photon Clarinets, and others) vary in price, depending on complexity, and have a bench rate of similar specialty repair or design professionals. For the unpainted "bare bends," I offer prices comparable to what I see other bent instruments sell for online; I do this to be competitive yet not undersell others while answering these requests.

Experience in a field is of utmost importance. This is especially true in bending, because outlaw electronics must be charted for a long while before becoming familiar enough to design fail-safe plans around. Untested bends often equal burn-outs, and burn-outs equal customer complaints: "No way! It lasted only two weeks!" Same is true of instruments that crash often or eat batteries. Back to the drawing board.

Sales Policies and Guarantees

I have two general sales policies. The first is for the instruments I design myself, entirely from scratch. These come with a one-year free service/replacement policy and a lifetime repair guarantee, as long as parts are available and I'm in business and able.

The second is for bent instruments using other manufacturers' circuits (as in the Speak & Spell); although I will try to repair an instrument for its lifetime, my free replacement warranty is for the first 90 days. Experience tells me that if the unit is going to go bad post-bending, it will happen within this period. After that I'll repair it at bench rates for as long as parts are available, for the original purchaser.

There is a sentiment within the bending camps that if it fries, it fries, and you're out of luck, "because I didn't manufacture the original circuit in the first place." I feel, however, that my choice in target instruments, my 38 years of bending, and my attention to fine prototyping should be able to culminate in some kind of assurance that the buyer is getting as solid a deal as the field will allow. Touchy subject, really—*should* the bender have to warranty the electronics of an antique Texas Instruments circuit? I choose to, but I can certainly see why others do not.

Certain steps to safeguard the instrument (such as immediate reset upon crashing) are covered in the instrument manuals that all major instruments come with.

Follow-up

Remaining at service to the customer is important. I continue to answer questions as long as they arise, and also reduce prices to prior buyers. I also encourage buyers to become benders. No, not the wisest business decision. But I'm here, as always, primarily to expand the art (my sales history is directly in response to requests).

Building Your Circuit-Bending Workshop

The workshop of a wizard is a very special vehicle. As did Thomas Edison's lab, your workshop will become a discovery factory. In this part, you'll assemble the right tools to get the most out of your circuit-bending lab and light up your own Menlo Park.

Tools of the Trade

I'd bent the nail into a general question mark shape, the head of the nail at the bottom. Clamping the nail in a vise, head up, I went at it with a large file. My aim was to turn the head into a small triangle shape, a shape that would fit the equilateral triangular recess in the odd screws holding my next bending project closed.

After about 20 minutes of filing, removing the nail to test the fit, and filing again, I finally ended up with a tool that should remove these weird screws. Holding the hook of the question mark as a handle, I placed the filed nail head into a screw's triangular hole. Good fit. Feeling quite confident, I turned the new tool to loosen the oddball fastener and promptly stripped the nail head into a nice cylinder as it slipped around and around in the stubborn screw.

Knowing that I could have just gone out and bought a set of "security drivers" with assorted unusual heads would have been nice. Knowing that some security screws unscrew clockwise would have been even nicer.

Choosing Your Soldering System

Such surprises as the odd hardware mentioned previously keep life interesting. And advising against the unknown is, at best, tricky. But the other half of this equation—entering the job with the right tools to begin with—I've got covered. I'll start with your soldering iron.

I've used 'em all. The mini butane soldering irons are nice for electronic repairs in the field, away from electric power. Big soldering guns are fine for larger solder connections than you'll be doing here. Stand-alone soldering pencils are fine if you file down the tips to a small beveled point. But what I've found to be best, and what you'll be using here, is a low-wattage (25–40 watt) soldering "station." This consists of a soldering pencil, pencil rest, and tip-cleaning sponge. A small tip is a must. It's also nice to have an adjustable temperature station, one that allows you to adjust tip temperature, but as long as the unit is made for small electronic work, you'll be okay.

I use two Weller stations, the WTCPT (my main station) and the smaller WLC100. They're reasonably priced and both are workhorses (see Figure 5-1). Each comes with cleaning sponge and pencil (iron) rest.

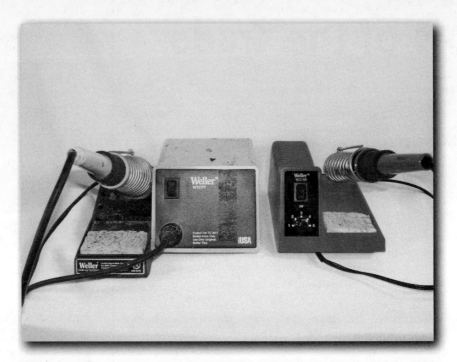

FIGURE 5-1: Weller soldering stations: the WTCPT on the left, the WLC100 on the right

Which Solder Should I Use?

I'm using the most common electronic solder available. This is the "60/40" rosin core solder. The only special requirement is its gauge (thickness). For delicate work, thinner is better, so buy a spool of rosin core 60/40 solder of .032 gauge.

So Many Tip Sizes to Choose From!

True. If you look at the Weller catalog you'll be amazed at all the choices. But here's what it comes down to again: tiny. The tip I use in the WTCPT is the Weller S-TPTA7. But others will work, too, if the shank is long (more than ½") and the tip is chisel or screwdriver type (flat on opposite sides) with blade tip, at best, no wider than ¹⁄₁₆" at the business end.

As of this writing, the tip supplied with the WLC100 is too wide for our needs. Be sure to get the ST6 tip to use with your WLC100. The WLC100 is very reasonably priced (about $40 as of this writing).

Avoid tips that are not chisel or screwdriver tipped! Purely cylindrical, tapered-to-a-dart-type-point tips won't hold molten solder as well as chisel or screwdriver-shaped tips do.

Common Tools

The idea I stress here is to *always* buy high-quality tools. Blade hardness, jaw alignment, and hand comfort are all important in small electronics tools. You'll often see great tool quality and prices at yard sales, if you're lucky enough to find electronics tools. Short of this, visit your hardware store or shop online and get the good stuff.

Wire Strippers

The main wire you'll be using is very thin: solid core "wire wrap" wire of 25–30 gauge. Your stripper's ability to strip such a thin gauge well is crucial—you'll be doing this a lot. Look for the simple stripper with two scissors-like jaws (not the "automatic" type that strips "all gauges" by sticking all wires in the jaws).

The correct kind will have two flat blades, each with a series of semicircular notches in the blade edges, ranging in size from small to large, each to strip a separate gauge (down to 30 AWG). Be sure that there is no slack or play in the axle that joins the blade halves. Blade halves should be snug against each other at all times. Spring-return and cushioned handles are a nice addition (see Figure 5-2).

FIGURE 5-2: Wire strippers

Wire Clippers

Although your stripper will most likely have a clipper section to the blades, having a stand-alone clipper is a must for getting into tighter spaces than the stripper will allow. Choose a pliers-type clipper with forged and ground blades over the pressed sheet steel kind. Again, spring-return and cushioned handles are nice (see Figure 5-3).

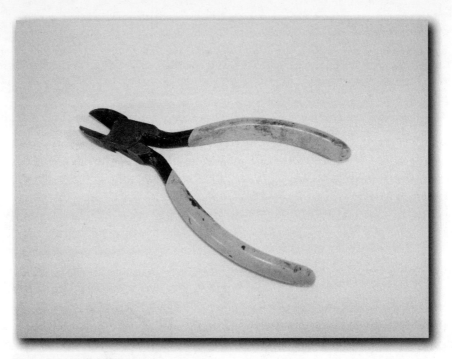

FIGURE 5-3: Wire clippers

Pliers

Everyone needs pliers. It's a general-use tool, and you're sure to find something to hold on to with it. Get a standard pliers as well as a "needle-nose" pliers, still thinking "small" for electronics (see Figure 5-4).

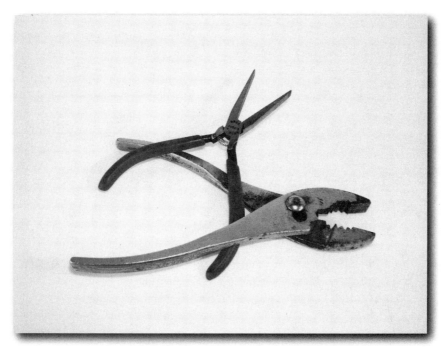

FIGURE 5-4: Small pliers, standard and flattened needle nose

Wrenches

We're talking crescent or box-end wrenches, smooth and chromed, like the mechanic's crescent wrenches but way smaller. These are used for precision tightening of panel-mount hardware (such as switches), but they come in handy elsewhere, too. You'll want to obtain a set that, in both standard and metric sizes, covers the range from about $3/4$" to $1/4$" (see Figure 5-5).

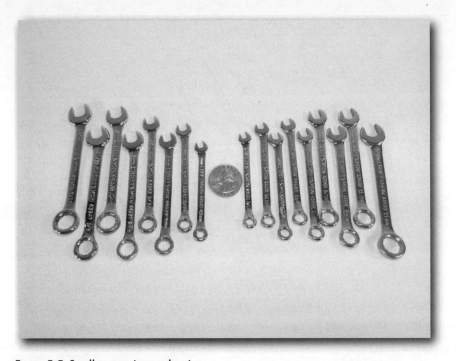

FIGURE 5-5: Small crescent wrench set

Note   Some Sears outlets carry sets of small crescent wrenches. They're also available individually, including designs incorporating ratchets in their box ends—like my $5/16$" for the thousands of panel nuts common in that size.

Drivers

By "drivers," I mean screwdrivers and their ilk. You'll need standard and Phillips drivers from medium sized on down to tiny. Buy a good "jewelers" driver set and you should be covered. You'll also need star (Torx) drivers in smaller sizes. There are many sets of drivers on the market that will supply you with all the standard, Phillips, and star drivers you'll ever need. Don't buy a set with one handle and many driver blades. Get a set with high-quality blades affixed within comfortable handles (see Figure 5-6).

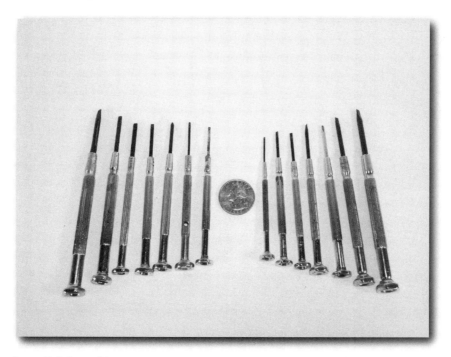

FIGURE 5-6: Screwdriver set

Note Really unusual screw heads are tackled with "security blades." My security driver set is by "SE" (part #7518SD), contains 30 bits, and is available through American Science and Surplus.

A socket driver looks like a screwdriver except that where the tip of the blade would be is a socket similar to the sockets in a ratchet wrench socket set. You'll also want a set of socket drivers, again in standard and metric, sized from ¾" on down to ¼". Be sure to get socket drivers for electronics, because these have *hollow shafts* instead of solid. The hollow shaft allows the handle of toggle switches and other components a recess while the wrench tightens their hardware (see Figure 5-7).

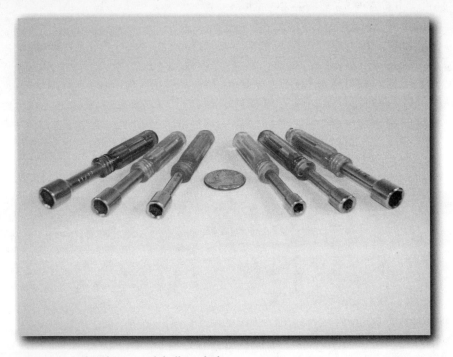

FIGURE 5-7: Socket drivers with hollow shafts

What about power drivers? A good-quality, power-drill-type screwdriver is a great addition to the bender's bench. My RYOBI 7.2-volt power driver has an adjustable clutch (to keep from overdriving), has good balance, holds a charge well, and is one of a few mid-to-low-priced power drivers available (be sure to buy a set of long bits–2" to 3"—if your power driver is not supplied with such). However, the emphasis in this book will be on hand tools, because most tool applications here are hand tool only, and because practice with hand tools of all types will hone bending skills and better familiarize the bender with the materials at hand.

Drills

Dremel drill sets have come a long way in the past twenty years (see Figure 5-8).

FIGURE 5-8: Dremel drill with adjustable chuck and LED illumination

You'll be using a Dremel drill to drill all pilot holes for component mounting. Get a good package that has an adjustable chuck (instead of collets) and includes a set of bits. If possible, buy the set that includes a flexi-shaft as well as the LED illuminator. If the set does not include "burr" bits, buy some separately in sizes from ⅜" on down (see Figure 5-9).

FIGURE 5-9: Burr bits

Although the Dremel is a must-have, there are other drills I use quite frequently (but are not needed for the usual bending projects and certainly none in this book). First is a drill press. If you need to use a hole saw to create a 3"-diameter hole in a case for speaker mounting, the press will come in very handy. The same goes for using larger bits too big to mount in the Dremel chuck. Having an electric hand drill is a good idea, too.

Next is a ceiling-mounted motor connected to a flexi-shaft. This is a high-torque machine capable, even with large burr bit mounted, of bringing tiny pilot holes up to size, production-line style, without ever asking for a coffee break.

Not-So-Common Tools (But You'll Love 'em)

A good tool can get you out of a bad spot. Here we'll look at a few items that are either a must or highly recommended.

Hand Bore

Inept gestures? No. We're talking about boring holes. A hand bore is a hand tool used to enlarge holes, such as the pilot holes drilled for component mounting. Bores have long, tapered, metal blades and often come in sets of two. The first will taper from about ⅛" up to ¾". The second then takes over, tapering from ¾" up to more than an inch. This is a must-have for general work as well as for making precise holes for mounting components larger than drill bit diameters. Be sure to have one (see Figure 5-10).

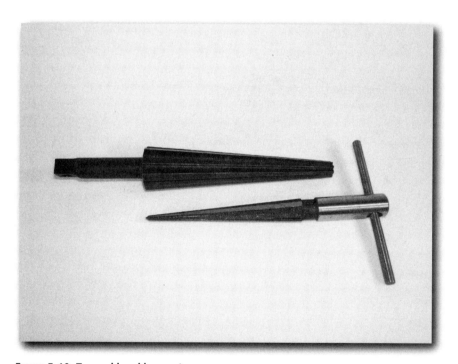

FIGURE 5-10: Tapered hand bore set

De-burrer

Not designed to warm you up (how could I resist?), the de-burrer is another hand tool I strongly suggest you get hold of. After using either a drill bit or the hand bore, you'll find that the edges of the hole are often rough. These "burrs" of plastic are quickly removed with a de-burrer, making for cleaner control mounting and much smoother, quicker assembly. If this is a difficult tool to find where you live, you can buy a de-burring bit made for a drill press and mount it on a handle of your own design. These bits taper up very quickly from a sharp point to a width of about 1" and are also referred to as counter-sinks (see Figure 5-11).

FIGURE 5-11: De-burrer

Rifflers

Rifflers are long, tapered files with variously shaped curving blades. Of oriental design, these files are great for precisely removing material from the workpiece. Because you will occasionally need to modify plastic cases, small files come in handy. These are not a must-have, but do consider them for fine work (see Figure 5-12).

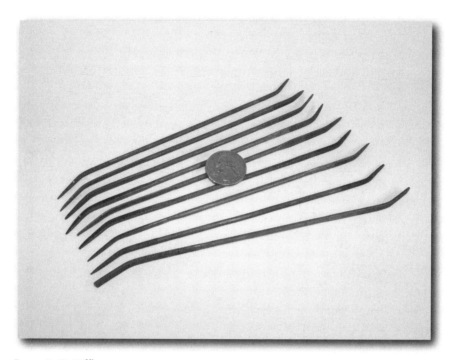

FIGURE 5-12: Rifflers

Glues

Not all components come with panel-mount hardware. And there are tons of other jobs that need one adhesive or another. Because we're dealing mostly with metals and plastics here, "white" glues are out of the question (Elmer's and other paper/wood glues need a porous surface for proper bonding). This fact brings us up to the table of elements a little further.

Hot-Melt Glue

Although strand-prone and potentially messy, hot-melt glue's ability to set quickly into a semi-solid mass has definite benefits. Your choice here is, essentially, among large and small glue guns. My suggestion is to hang with the standard hot-melt gun because the glue sticks are also standard and easily found. The smaller gun, available at hobby shops, is nice, too, as long as you keep the smaller-diameter glue rods in stock (see Figure 5-13).

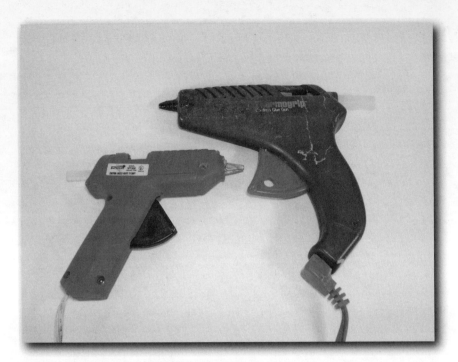

FIGURE 5-13: Standard and small hot-melt glue guns

With the larger gun comes the option to purchase extra-long glue sticks. But remember that a semi-used stick is glued into the gun as soon as it cools, and the glue gun is now two-and-a-half feet long. Yes, a utility knife can solve that problem, and even chopping long sticks down to size upon purchase can save a buck. But I "stick" with the shorter breed most of the time.

You'll see a few glue variants. General purpose is fine, carvable, removable (with a small chisel blade), sets in about 3–4 minutes, and is what you'll most likely see at the stores.

As to the gun itself? The most important feature is a power light (unless you enjoy the smell of charred workbench waking you at three in the morning). These are getting harder to find in the rush to manufacture items as cheaply as possible, come-what-may. That penny saved may be a workshop burned. Get one with a light if you can.

Silicon Adhesives

Silicon adhesives are everywhere now, even holding great slabs of stone on the façades of buildings in place of more traditional cement mortars. At least for the time being.

Silicon glues are stickier than hot-melt. That's good in an instrument that will be flexed, not uncommon in the plastic-housed circuits you'll be dealing with. Silicon is also removable using small-bladed tools. But although it skins over quickly, it still takes a long while to cure to full bonding strength.

Silicone is a "broad spectrum" adhesive, capable of bonding porous and nonporous materials alike. Crafter's Goop and other specialized versions are available, some setting more quickly than others. Having a tube at hand is often convenient.

Epoxies

Epoxy comes in handy. Slower-setting epoxy is generally higher in strength. But "5-minute" epoxy will usually work fine. It's more than strong enough for our needs and cures fast enough that you can hold parts in location as it sets. Roughen plastics before applying epoxy (cross-hatching with a pointy object works well).

There are also epoxy putties that mix together prior to setting, just like the liquids. These are sculptable as they set up, and can be used where reinforcement is needed in mounting larger or unusually shaped parts.

Mask Up!

We'll be battling particulate matter: It wants in, we want it out. Having the right respirator at hand is only common sense (see Figure 5-14).

If you intend to spray paint your instruments, be sure to wear the appropriate mask. The same with drilling at higher speeds or any activity that produces airborne particulates. *Always* mask up!

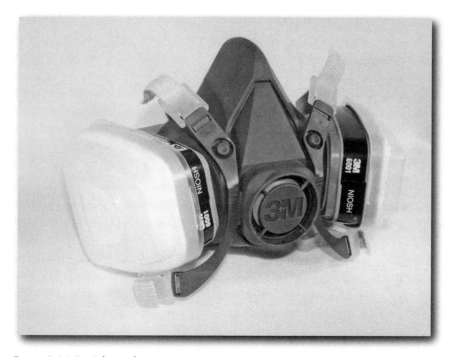

FIGURE **5-14: Particle mask**

Your Electronic Parts Supply

When I began to bend circuits, I had no real idea as to what the components actually did. I would choose one resistor over the other because it had a lavender band instead of less exciting colors. I'd choose a component because it had a weird shape, not knowing or caring whether it was a resistor or capacitor, a coil, transistor, or photocell. All I knew was that it might change the sound in an interesting way when included in the circuit-bending path. So I tried it.

Ridiculous, I suppose, and downright foolhardy if you're dealing with anything other than very low voltage. But had I not, we wouldn't be here today, and the monkey would still be at the typewriter (he said, checking his knuckles for hair).

What Are All These Things?

Let's take a look at the menagerie of components you're most likely to use in the projects ahead, and the most common in bending. Just about all of them will be used in the middle of the new circuit-bending paths you'll discover, so don't get excited about not knowing what to do with this or that. Read on and get comfortable. This stuff is really cool (see Figure 6-1).

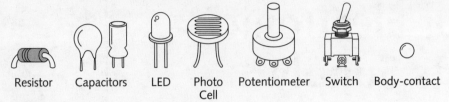

Resistor   Capacitors   LED   Photo   Potentiometer   Switch   Body-contact
Cell

FIGURE 6-1: Circuit-bending's most common components

Tying It All Together: Electricity

Look, we don't really understand everything about electricity yet. Arguments still abound. I'm certainly not going to pretend that I do. And as promised, you don't need to understand it, either. But because this is the magic juice that brings all our components to life, let's at least see what it's up to.

The usual analogy of electricity is that of a water system. Just as water flows through a pipe, electricity, made up of electrons, flows through a wire. Lots of electrons.

In fact, just as water has a current, so does electricity. The current of electricity is judged by how many electrons pass a certain point within a given time. The unit of measurement here is an *ampere*.

Not that it's important to us, but because people are impressed by big numbers: 6,250,000,000,000,000,000. That's how many electrons have to pass a specified point within one second to equal 1 ampere of current. Not pokey at all, electrons move at just about the speed of light.

Electricity, again as with water, follows the path of least resistance. Whereas water seeks valleys, electricity seeks metals. Just as using a hose siphons water from higher to lower pressure, metal wire "siphons" electrons from higher to lower "potential." What's important in each case is that a flow is established, and that we can use that flow.

When you circuit-bend, you'll mostly be soldering individual wires between points on a circuit (see Figure 6-2).

All these electrons I'm talking about will flow through these wires like water through a hose, from one point on the circuit to another.

If you cut the wire in the middle and solder, for example, a resistor in that gap, the electricity will then have to flow through the resistor to get where it's going. The components soldered into the middle of the wire will, you hope, change the sound of the instrument in an interesting way. So let's run down the list of your best bet for electronic parts to keep at hand, and talk about what they actually do.

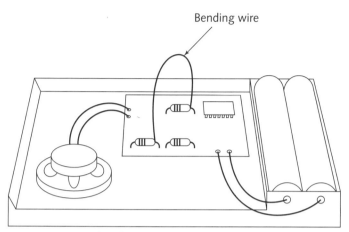

Bending wire

FIGURE 6-2: Circuit-bending wire soldered in place

Switches

So, you turn your circuit on and discover that if you touch one end of a wire to this spot, and the other end to that spot, you get a sound like someone playing a single flute note, such as B-flat, in an empty aircraft hanger.

Nice echo, good tone, but flute notes can get tiring. You need to be able to turn the spacey flute note on and off at will. You could unsolder one end of the wire that was added to create the note. That would "break" the circuit, ending the flute note. And you could solder it back in place when you wanted to hear it again. But what a hassle. There's gotta be an easier way!

Toggle Switches

All switches either "make" (connect) or "break" (disconnect) a circuit. Meet the mini toggle switch. If you solder the switch in the middle of the wire that you soldered in place previously, you can then turn the flute on and off just by flipping the switch.

If the toggle switch you're working with has only two soldering lugs on the back, you'd snip the flute sound wire in the middle and solder the two wire ends to either of the two lugs, with each wire end connected to a separate lug (see Figure 6-3).

If the toggle has three soldering lugs, you still use only two. In this case you'd solder one wire end to the switch's middle lug and the other wire end to the bottom of the switch's remaining two lugs. The bottom depends, of course, on how you mount the switch, but with one wire to the middle lug and one to the bottom lug the switch will work the same as the light switch on your wall: To turn it on, you flip it to the UP position (see Figure 6-4).

If you instead solder one wire to the middle but then solder the remaining wire to the upper in place of lower lug, no harm is done. It's just that the switch will then operate in reverse: To turn the flute sound on, you have to turn the switch "off" (downward instead of upward). The flute still sounds the same, but your friends will have no remaining doubts.

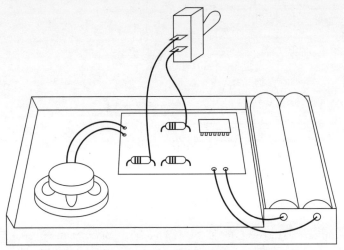

FIGURE 6-3: Circuit-bending wire with two-pole on/off switch

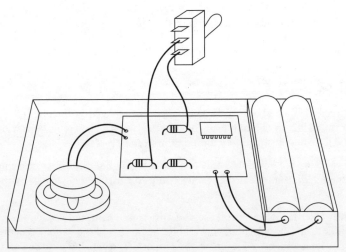

FIGURE 6-4: Circuit-bending wire with three-pole on/off switch

Pushbutton Switches

Doorbells are pushbutton switches. So are the piano keys on your synth. Whereas the toggle switch "makes" (connects) a circuit by flipping a switch handle, here you just push the button to connect the circuit.

Simple pushbuttons have only two soldering lugs, and either wire can go to either lug. Such a switch is called a "normally open" switch because the switch does not connect, or close, the circuit until pressed—the circuit is normally open until closed by the switch (see Figure 6-5).

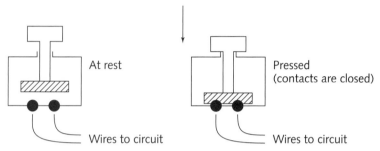

At rest

Pressed
(contacts are closed)

Wires to circuit

Wires to circuit

FIGURE 6-5: A normally open pushbutton switch in action

This kind of switch is just what you need to turn on a circuit for a moment, as in the flute example. Press the switch and you'll hear the flute as long as the button is held down. But there are times you'll want to push a button to turn a circuit off, such as turning the battery power off to re-initialize a crashed circuit. Such a pushbutton is called a "normally closed" switch because, as opposed to the normally open version I just described, pressing it *dis*connects, rather than connects, the circuit (see Figure 6-6).

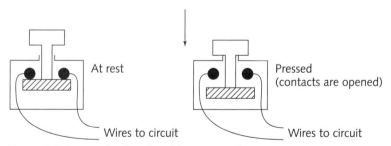

At rest

Pressed
(contacts are opened)

Wires to circuit

Wires to circuit

FIGURE 6-6: A normally closed pushbutton switch in action

Pressing a normally open switch closes the circuit (the same as soldering the end of the wire back in place on the board). Pressing a normally closed switch opens the circuit (the same as de-soldering the wire again and removing it from the board). In other words, switches save a lot of solder.

These two examples are of "momentary" pushbutton switches. Their effect is momentary, or remains as long as you press the button.

There are also pushbutton switches that connect or disconnect a circuit every time you push and let go. If your doorbell were connected to a switch like this, the UPS guy would press it once (the doorbell buzzer is now on and the live recording of your dripping sink is ruined), and then walk away thinking you weren't there, doorbell still buzzing (you go downstairs to push the button again; buzzer stops, sink overflows). These are referred to as "push on, push off" switches. You don't need 'em for this book's projects, but you can see that they're lots of fun.

So, you'll be stocking up on *momentary* mini pushbutton switches of both types, normally open (N.O. appears on the packaging) and normally closed (N.C., likewise).

Rotary, Micros, and Motion-Sensitive

Though the projects in this book don't require these variants, all can be used in creative circuit-bent designs. Rotary switches can send the spacey flute circuit through numerous new components, such as various resistors or capacitors, just by turning its dial. A micro switch's "feeler" arms can be extended to become keyboard-like keys, and motion-sensitive switches allow you to move an instrument around in space to play it (see Figure 6-7).

Rolling Your Own!

Switches are very simple. They just connect or disconnect two wires, usually. If you buy flat, brass-rod stock, you can fashion your own "reed" switches (no, I can't claim the honor—reed as in "an elastic tongue of metal").

From brass stock (available at hobby shops), cut two reeds of the same length and solder a wire to an end of each. Wrap ½"-wide masking tape around one of the reeds just past the solder until the tape builds up to ⅛" thickness. Hold the remaining reed parallel to the first and wrap the entire unit with tape a few more times. Trim the tape and coat the tape junction with a quick-setting glue such as five-minute epoxy or hot-melt. Any pressure on either reed toward the other will now close the switch and connect the circuit running through this switch (see Figure 6-8).

Custom switches like these are nice for unusual applications involving limited space or unordinary, hand-fashioned switch buttons (see the section about Vox Insecta in Chapter 11).

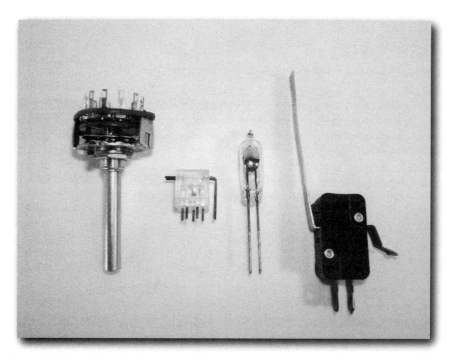

FIGURE 6-7: Specialty switches: rotary, ball, mercury, and micro-switch

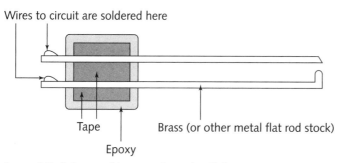

FIGURE 6-8: Cutaway of homemade reed switch

Resistors

Resistors are all those little cylinders with colored bands toward each end that are lying flat to the circuit board. Their job is to resist current flowing through them.

Inside, they're pretty boring. The usual carbon resistor is essentially a mixture of powdered carbon and electronically neutral glue. Carbon conducts electricity pretty well. The more carbon added to the glue, the better the resistor conducts the electrons flowing through it. But no matter how much carbon is in there, a resistor *always* resists electrical current.

If in place of the switch you put a resistor in the middle of the flute circuit wire, what will happen? The current flowing through the wire will be resisted and the flute sound will probably be lowered in pitch (see Figure 6-9).

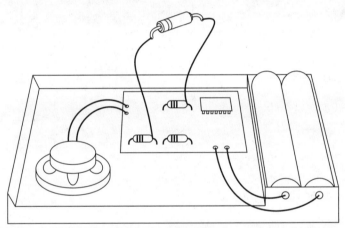

FIGURE 6-9: Circuit-bending wire with resistor

If you then solder the switch back somewhere into this wire, on either side of the resistor, you'll be able to turn on and off the lower-pitched flute sound whenever you like (see Figure 6-10).

Other uses for resistors in circuit-bending are lowering the current to "cool down" audio outputs that are too loud, and to lower current to avoid burning out the LEDs that you'll be using as pilot lights and envelope lamps. Don't worry, I cover this in detail later. It's all very simple stuff.

But which resistor? Because there are many resistance levels offered by different resistors, as indicated on them by the color-coded bands, the easiest way to experiment with resistors is to use a resistor substitution wheel. This is a small device loaded with resistors inside. You simply clip its two leads into the circuit with which you want to experiment and turn the dial (see Figure 6-11).

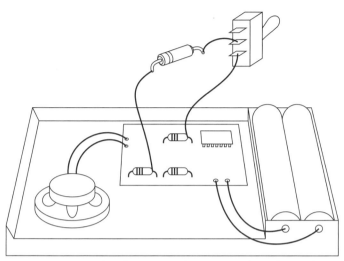

FIGURE 6-10: Circuit-bending wire with resistor and toggle switch

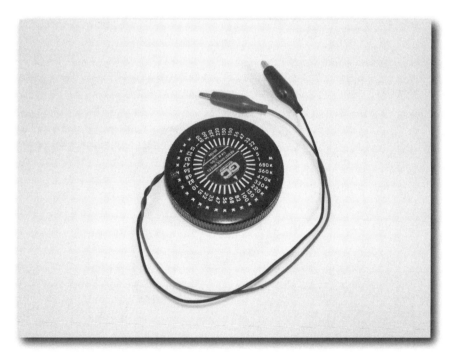

FIGURE 6-11: A resistor substitution wheel

As you rotate the dial, each click of the wheel connects a different internal resistor between the leads and allows you to see or hear the effect of that resistor on the circuit. Resistor values are printed on the case just outside the dial so that you know which resistor is connected as you turn the wheel.

When you hear or see the result you want, stop turning the wheel and look at the resistor value that the dial is pointing to. If it says "500K," you know you need to solder a 500k resistor into the circuit to get the same effect. Resistor substitution wheels are very handy to have around, and one should be part of your bending kit.

Note They're getting harder to find, but a Web search will usually uncover a wheel by GC Electronics, part #20-104.

Potentiometers

Potentiometers, called "pots" for short, are also resistors. The difference is that their resistance is variable. The volume knob on your stereo is a pot. As does the resistance substitution wheel mentioned previously, they adjust resistance as they're turned. But unlike the wheel, their change is smooth instead of stepped across their range of resistance.

Inside a pot is a resistance strip whose ends are connected to the outside soldering lugs on the pot's case. The remaining lug, in the center, is attached to the pot's "wiper." Turning the pot's shaft rotates the wiper along the length of the resistance strip.

The reason that resistance is variable is that the wiper position allows electricity to pass through more or less of the resistance strip, depending upon how much the shaft is turned. The more of the strip the circuit has to flow through, the more resistance is encountered (see Figure 6-12).

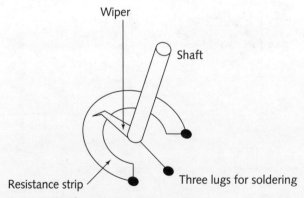

Wiper

Shaft

Resistance strip

Three lugs for soldering

Figure 6-12: The potentiometer in action

If you solder a pot into the middle of your flute circuit, you'll probably be able to adjust its pitch up and down as the pot's shaft is turned (see Figure 6-13).

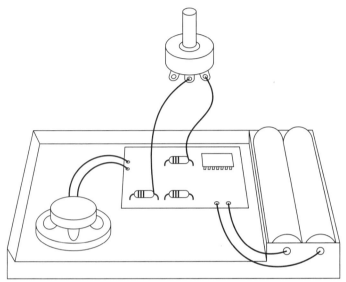

FIGURE 6-13: Circuit-bending wire with potentiometer

Pretty cool, especially because I'm using the bent flute sound only as an example, and you'll likely be working with much stranger sounds, sounds whose characteristics will be changed in often amazing ways as the pot's shaft is turned.

Pots usually have three soldering lugs on the back. You'll always use the center and one of the outside lugs to solder to. Which outside lug you choose depends on whether you want the pot's effect to happen as you turn the shaft clockwise or counterclockwise.

For example, if the volume control potentiometer on your stereo had its outside lugs wired in reverse, the pot would operate in reverse: The volume would go down as you turned it up. For volume and pitch controls, you'll probably want to stick to the usual wiring scheme, in which each goes up when the pot's shaft is turned clockwise. If the reverse happens in a design of yours, just reverse the wiring of the outside lugs.

You'll have tons of use for pots in circuit-bending, so have a full range of pots at hand, from 5K to 10M and all the steps in between (shop surplus). Most common are those with 1" diameter bodies and ¼" shafts. Panel-mount hardware (washer and nut) should be included (see Figure 6-14).

FIGURE **6-14: Assorted potentiometers**

The golden rule for pots in bending is: *Substitute!* When you find a good spot for a pot, try several pots in that circuit, ones ranging, again, from low resistance (5K) to high (10M).

Trimmers

Not for panel mounting, trimmer pots are to be used within circuitry to create precise resistance values. Typically tiny, these potentiometers are used to adjust voltage thresholds for circuit stability (see Figure 6-15).

Linear and Nonlinear?

You'll have a choice at times between "linear taper" and "audio taper" pots. The ratio of resistance is changed disproportionally with audio taper pots, whereas linear pots change resistance uniformly across their range. As you collect pots, you'll be able to try both types and see how their response differs in various applications. Because bending is an art of personal taste and experimentation, I won't recommend one over the other. Experiment!

FIGURE 6-15: Assorted trimmers

Capacitors

Resistors' best buddies are often capacitors, caught together with their pants down time and time again in the "r/c" combination. Here a resistor is used as a current adjuster to regulate electrical current flowing into a capacitor, whose job it is to store it. A capacitor, or "cap," not only stores electricity but also releases it as soon as it builds up enough of a charge to break down its internal discharge threshold or until it's otherwise triggered to discharge for some more entertaining reason.

For example, the electronic flash on your camera is capacitor based. Because the flash needs far more momentary energy than the battery can deliver all by itself, the battery therefore is used to charge a cap that in turn stores electricity, allowing it to build up. This built-up energy is finally discharged through the gas in your camera's flash tube, the holiday gang all get demonic red eye, and the capacitor begins slowly filling up again, like an old water tower fed by a garden hose.

In this book's projects, you will instead be using dinky caps for tone adjustments here and there, often spliced into the middle of the flute wire you've been bending back and forth all this time. Soldering a small capacitor in the middle of your flute-sounding bend might give the tone of the flute a sharper edge, taking the smooth sine wave of the flute and pushing it toward more of a triangle or square wave-form, making it sound more like a clarinet or even a cello. The capacitor is doing the same thing that it did with the flash example, only much faster and on a smaller scale.

The electrical current that passes through the wire that created the fine flute sound enters the capacitor smoothly and in a steady flow, par usual. But the cap now does its charge/discharge thing to this current, and, like a balloon against bike spokes, the outcome is noticeable.

You'll be fine with an assortment of low-voltage "disc" as well as electrolytic capacitors, best priced in variety packs as well as in bulk as surplus (see Figure 6-16).

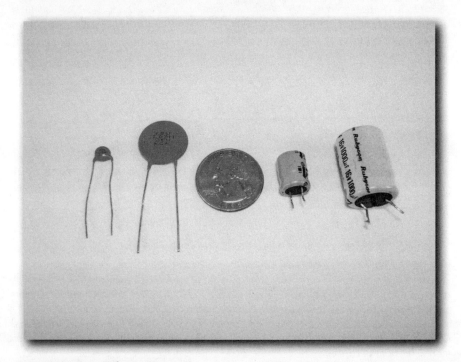

FIGURE 6-16: Assorted capacitors

Electrolytics Can ZAP!

Larger capacitors of the "electrolytic" type, as I touched on in the flash tube example, are not to be touched on, literally. At least not without discharging them first. You won't be dealing with these, most likely, but persons who are into electronics should be aware that larger electrolytic capacitors can maintain a big charge for a long time after being removed from an active circuit.

Sounds like a battery, the way it's maintaining a voltage like that, right? Exactly. In fact, electrolytic caps are commonly used as "batteries" in power supply back-up systems. They kick in when the main power supply is interrupted and, in this case, slowly release their charge as would a battery.

A 1K or larger resistor can be placed between the leads of a suspect capacitor to discharge it. Using a needle-nose pliers with insulated handles to hold one end of the resistor, place the body of the resistor between the leads of the capacitor. If charged, the capacitor will discharge from pole to pole through the resistor and then be safe to handle. No, shorting the cap's leads without a load (the resistor) in place is not a safe option. The fast discharge so produced can explode the cap or cause other nastiness.

Ask your parts dealer to point out large capacitors to you so that you'll recognize them and then steer clear if you see a group of 'em wandering down the dark alley where you're Dumpster diving. Unless, of course, you're carrying your 1K resistor with you, which you can swiftly draw from your shirt pocket and spin threateningly between thumb and forefinger, the flickering yellow of the alley's street lamp glinting off its bent leads.

Some capacitors are "polarized" components. As do batteries, these have both positive and negative poles (usually well marked on the case). In critical circuit design, these poles must be respected. In your work here, this won't be too critical an issue. You're just substituting components for interesting sounds, come what may, and you're specifically exploring *out-of-theory* here.

But as you build kits or design your own circuits later from scratch, be aware of capacitor polarity as well as voltage rating (never use a cap of a lesser voltage than it will be exposed to in the live circuit—again, not a worry to benders within the routine work outlined in this book).

LEDs

Semiconductors are glowing all over your circuits everywhere. If you could see infrared light, otherwise dark circuits would look like well-illuminated miniature science fiction cities with ICs, transistors, and diodes all throbbing with eerie, scintillating light. Light-emitting diodes, or LEDs, take advantage of this beguiling aspect of semiconductors and tune the emitted light to the visual spectrum (as well as fine-tuned, powerful but still invisible infrared for data transmission).

Light-emitting diodes are *fun*. They're very efficient light sources, meaning that they draw little current to operate. They're bright at low voltages and don't drain batteries nearly as fast as the tungsten panel lights they've replaced. (The story's not quite that simple—see tungsten bulbs in Appendix C.) They're inexpensive and now come in just about all colors (white LEDs can be painted any transparent or translucent color as well). They're easy to work with and will reveal to you all kinds of activity not otherwise visible within the circuit you're working on. In this book, you'll be using LEDs as indicator lamps for several types of circuit functions (see Figure 6-17).

FIGURE 6-17: Assorted LEDs

Polarity?

Still, they're diodes. And a diode's primary function in circuitry is to act like a bouncer at the bar. Make trouble and you're out, and an electron moving against the flow in most circuits is about as acceptable as Tom Waits drinking an entire bottle of scotch at the pub and then insisting on leaving through the ladies' room.

A diode will allow current to flow through it in one direction only. Apply current in the other direction (such as turn the battery around) and the diode will kill the circuit. You're not getting through. So again, you're dealing with a polarized component that must be wired correctly to operate. No problem—you'll see how this works a little later.

For now, you just want to collect a bunch of LEDs in all styles, from run-of-the-mill 1.5-volt types to high-brightness, 3-or-higher-volt models. It's best to always have high-brightness white, blue, and red LEDs in your supplies because these will respond better to certain circuit conditions than will the lower-voltage kind.

Photocells

Like your own eye, photocells are sensitive to light. Being sensitive to light means that you can see. You can give your bent instruments vision, too, with cadmium sulfide photocells. This is very cool.

A photo-sensitive instrument can be played without touching, just by waving your hands in space above its photocells. Photo-sensitive instruments are also stand-alone kinetic instruments, capable of musically interpreting the ambient light and shadows of a passing day.

Photocells are also called photo resistors. This is because, as with potentiometers, they are able to vary resistance as electrical current passes through them. In the example of a pot, as covered previously, you turn the shaft to implement this resistance change. With a photocell, all you have to do is cast a shadow onto the cell to get the same change (see Figure 6-18).

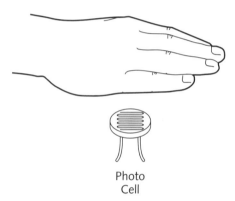

Photo
Cell

FIGURE 6-18: Waving a hand over a cadmium
sulfide cell changes its resistance.

What you might remember about photocells is that they're sleepy in the dark and excited in the light, like some of your friends. When in the dark and sleepy, they just lie there and ignore the electrons wanting to get through. But in the daylight, the photocells awake and, groggy from all the bent music the night before, slowly begin to let the by now riled-up electrons on by.

Considering low electronic social status (even a lead [carbon] pencil line drawn on paper is a resistor), miserable job requirements (keep rowdy electrons from going where they want), low benefits (maybe get turned off once in a while) and low retail value (photocells are even used as craft supplies for eyes on bunny rabbits), photo resistors often have it pretty rough. So use them often and let them do great things!

Recap: From Flute to Alien Instrument in Three Easy Steps

So, first you found a nice flute sound when you touched a wire from one random point on the target circuit to another random point. Next you soldered a toggle switch in the middle of the wire so that you could turn the flute sound on and off. After that you experimented with resistors, potentiometers, and photo resistors to adjust pitch, and then soldered your choice in-line with the toggle switch in the middle of the wire (see Figure 6-19).

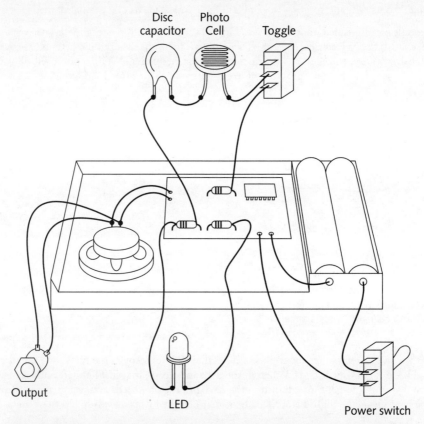

Disc capacitor   Photo Cell   Toggle

Output

LED

Power switch

FIGURE 6-19: Circuit with photo cell, capacitor, bend switch, power switch, LED, and line output

You decided on a capacitor to make the flute sound sharper, more like a wild cat's howl. You used a photo resistor so that you could get this alien cat to sing up and down a frequency range from cat-fight highs to low, window-rattling purrs. And you got these sounds by just waving a single hand in space, playing this electric cat eye with shadows.

Maybe you even used another wire, this one with an LED soldered in the middle, to bridge circuit points to discover ways to make it flicker as the voices play. You could add a line output coming from the speaker (covered in detail later) and stop right there, mount everything in a small case, and have a portable electronic cat with eye, voice switch, flickering LED, and an output for feeding into an amp. You could add a power switch too simply by adding another toggle switch to interrupt the battery power (see Figure 6-20).

FIGURE 6-20: Typical bent circuit in new housing

Wire It Up!

So far, I've discussed the cool stuff to put between the wires. But what about the wire itself? *Very* good question.

Stranded vs. Solid Core

If you look at both a section of solid-core and a section of multistrand wire side by side, you won't be able to tell the difference. But inside the plastic insulation, there's a big difference. Whereas solid core is constructed of a single metal wire inside its insulation, stranded wire has many thin wires inside.

For point-to-point wiring on a circuit board, and in almost all the wiring you'll be doing in this book's projects—board-to-switch or to other panel-mounted components—you'll be using wire that was not meant to be soldered at all. This is the wire you bought your wire stripper to handle: 25- to 30-gauge (AWG), solid-core "wire-wrap" wire (see Figure 6-21).

FIGURE 6-21: Wire-wrap wire is thin!

Be sure it says "wire wrap" if you're not familiar with this wire already.

This is a wire meant actually for solderless prototyping, for which a special tool is used to strip and wrap the exposed wire around electronic component leads. But it's great for circuit-bending, and for a number of reasons.

Mostly, wire-wrap wire is good for our needs because it's very thin. This means that it will get into tight places (such as under resistor leads already soldered to the board). And because little metal is involved in thin wire, it absorbs heat quickly and will therefore solder very fast.

Real wire-wrap wire is a "silver content" wire, actually containing silver. This does aid in soldering, but regular (nonsilver) "wire-wrap" wire (same gauge, pre-tinned, probably won't say "wire wrap" on it) solders equally fast. Fast is important when you're soldering to heat-sensitive items such as transistors, IC pins, and diodes.

Beyond this, wire-wrap wire is very easy to strip, its insulation is usually high quality, the inner wire is always solder friendly (not all bare wire is), and, c'mon, it comes in marvelous colors.

You'll always find wire-wrap wire super cheap at the walk-through surplus warehouses (such as Mendelson's in Dayton, Ohio). I buy 500' rolls a few at a time if I like the gauge and color of the insulation. The standard 50' rolls are available at Radio Shack. It's best to have three or more colors at hand so that you can devote single colors to separate circuit areas. This way, tracing your new circuits later is much easier.

As mentioned, any 25–30 gauge, insulated solid-core wire will work. It does not have to be authentic silver content wire wrap, but it should still be shiny silver in appearance under the colored plastic sheathing. The only caution: Try not to flex these wires after they're soldered in place. Too much flexing will break any solid-core wire at its connection points.

Stranded vs. Solid Core: Who Wins?

You should also have at hand a spool of multistrand wire. Why? Each will conduct electricity as well as the other. But solid-core wire, like wire wrap, will not take as much flexing as multistrand will before it breaks.

Say that you have an instrument you're working on for which you have to open the case to replace the batteries. But you've added switches to the side of the case opposite the half where the circuit board is mounted. Every time you open the case, these added wires are flexed. Although wire-wrap wire is fine from point-to-point on the board itself, connections that will be flexed, those going from the board to the switches on the removable half of the case, should be made with stranded wire of a heavier gauge (noncritical; think uncooked spaghetti for overall thickness, or thereabouts).

Tip   Here's a nice trick. If you're unsure whether the wire you're looking at is solid core or stranded, just bend it. If it holds the bend, it's solid. If it tries to bend back, it's stranded.

Did I Miss the Bus Wire?

Much wire is, as just discussed, insulated with an outer plastic covering. This is done so that when these wires touch each other, or different parts of the circuit, they don't make unwanted connections and really mess things up.

But there are many cases in which you'll be wiring things in series that need a common wire to connect to many different points. This is called a "bus" wire because it buses the same electrical current to many places. Often a "ground" wire in traditional electronics (in which it connects many components needing to be attached to the negative side of the circuit—don't worry about it), for this book it will usually be a "common system wire" connecting, say, a bunch of switches all to one point. So what?

Well, removing insulation is a hassle, no matter how easy it is. Bus wire is a solid wire with no insulation at all. And if one wire needs to go to many different components on a control panel, bus wire might be the way to go as long at it won't touch other circuit areas (see Figure 6-22).

Bus wire comes in various gauges (you'll see 26 to 22AWG in surplus on huge spools). The usual product so named at Radio Shack will be fine.

So, have at hand several colors of wire-wrap wire, a spool or two of insulated multistrand wire, and a spool of bus wire.

To circuit

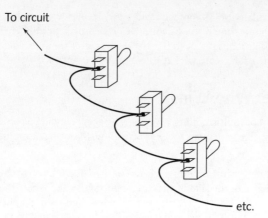

etc.

FIGURE 6-22: Bus wire used to connect common
contact in a row of switches

Finally, you'll need to have some shielded audio cable around. This is the cable your phono
cords and guitar cords are made from (see Figure 6-23).

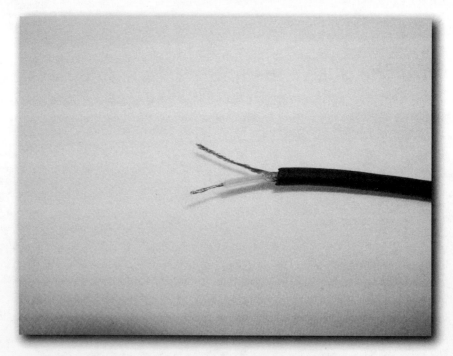

FIGURE 6-23: Shielded cable with inner wires exposed and stripped

Shielded cable has one or two (sometimes more) multistrand, insulated wires running inside a braided outer wire "shield." Buy this in coils rather than wrapped on spools if possible (the spools can damage the flexibility of the cable).

Jacks of All Trades

Jacks are, of course, those hot-wired holes that get signals in and out of audio electronics. You'll be using jacks to help you create "line outputs" with which you can run your bent instruments into an amplifier, a mixer, or an effects unit. Among the dozens of jack types available, only a few will be needed for most bending work.

All the projects in this book use mono jacks, and all mono jacks are two conductor. That is, they have two soldering lugs—a "hot" and a "common" or "ground." The "hot" lug connects to the part of the jack that touches the tip of the plug when inserted fully. And that leaves but one side remaining: the common or ground.

If the jacks are supplied without a diagram telling you which lug is which, just take a close look. Follow the section on the jack that would touch the tip of the inserted plug and then follow the metal contacts down to a soldering lug. If the connections on the hot side of things are foggy, study the other side and see where its metals connect. You should be able to trace one side or another to a lug and thereby be able to determine which lug is which for the jacks you'll be using in this book's projects.

You Know It and Love It—the "Guitar Jack"

Limited space is often a norm in circuit-bending. Guitar jacks, as on a guitar amplifier, accept a $\frac{1}{4}$"-diameter plug, exactly the thing on the ends of your guitar cord. Although $\frac{1}{4}$" isn't all that big in itself, the shaft of guitar plugs is pretty long, and the body of the jack you're plugging into takes up a lot of space as well (see Figure 6-24).

Too, guitar cords are often quite heavy. If the instrument itself is lightweight (another norm in bending), attaching a heavy guitar cord can nearly pull it off the table.

On the bright side, guitar plugs and jacks are sturdy, secure mechanically and electronically, and common. If the instrument you're working on won't be too lightweight, and if there's room for a guitar jack, there's no problem.

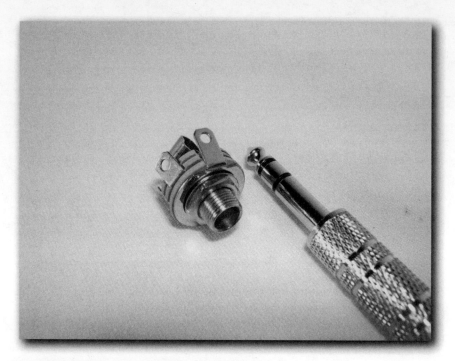

FIGURE 6-24: One-quarter-inch guitar jack and plug

RCA or "Phono Jacks"

These are the jacks you probably have on the back of your stereo system or stereo audio amp. These jacks and their plug counterpart take up much less space than the guitar jack and, if cared for, will provide good service (see Figure 6-25).

The problem with RCA jacks is that the panel-mount version's back-mounted hex nut can loosen over time, needing a repair. This happens only if you're not careful to pull and push plugs straight into and out of the jack. If you forget this simple rule and instead twist plugs in and out, as we're used to doing on the back of our stereos, the jack may eventually give you trouble.

On the bright side, as with guitar plugs, RCA jacks are sturdy, secure mechanically and electronically, and common. RCA "phono cords" are everywhere in audio work, and compatibility with mixers is built-in. And as I've said, they're space saving.

FIGURE 6-25: RCA jack and plug

Mini Jacks

The mini jacks I'm talking about here are the size of a headphone jack on a Walkman. These ¹/₈" jacks are tiny versions of the guitar jack and are in all ways similar, including the wiring. Smaller still than the RCA jack, for really tight spots they're problem solvers (see Figure 6-26).

On the down side, these jacks are frail. If you work with a ¹/₈" jack a lot, it may develop play in its mechanical connections. But then, all work and no play makes . . . oh, never mind.

Anyway, have at hand an assortment of these three jacks and you'll be well covered as you bend into the future.

FIGURE 6-26: One-eighth-inch mini-jack and plug

Battery Holders

We know 'em and hate 'em: battery holders. Although you won't need these to bend the instruments in this book, there will be times in the future of all benders when you'll want to remove a circuit from its original housing and move it into a larger case. The original, usually built-in, battery-holder is left behind in this transformation, requiring a replacement.

Battery holders are available in all sizes, holding everything from the single 9-volt rectangular battery to any multiple of the cylinder cells you'll be using. These can be bought to requirement as the need arises (see Figure 6-27).

Dummy Cells

What if you're replacing a battery compartment that held 3 "C" cells and you can buy only a replacement battery compartment that holds four? Making a dummy cell to take the place of the unneeded fourth cell's space is the answer.

Using a wooden dowel whose diameter is the same as or a little smaller than the batteries you'll be using, cut it to $\frac{1}{8}$" less than the usual battery length. Solder a thin washer to each end of a wire whose length is just a little longer then the dummy battery, terminal to terminal. Affix these washers to either end of the dummy using glue or double-sided tape and insert along with the other batteries in the new holder (see Figure 6-28).

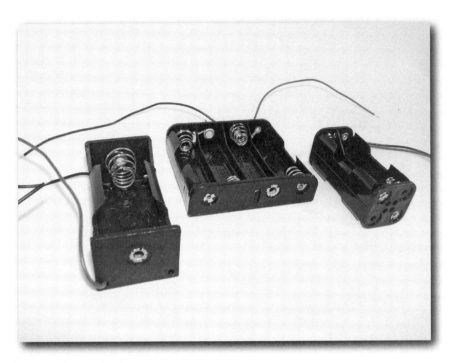

FIGURE 6-27: Assorted battery holders

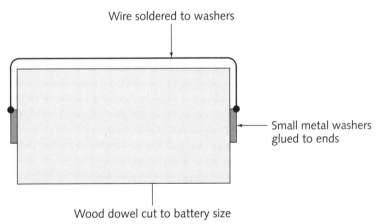

FIGURE 6-28: Making your dummy cells

Electricity will now flow around the dummy battery, following the washer-tipped wire. In this way the four-battery-holder now holds only three, and the voltage it's usually supposed to supply (6 volts) is dropped down to the 4.5 volts you need.

Tip Soldering a wire between the spring and opposing terminal of the extra battery chamber will also do the trick—if you don't melt the plastic and ruin the battery holder.

Electrical Tape: Not a Good Thing

Electrical tape, as in the elastic, black, sticky-edged puck of old, has seen its day. At least for circuit-benders.

The problem with the old tape is that it does not age gracefully, unless you think that oozing sticky tar all over fingers and wires is graceful. Need a second definition? How about dry, black flakes crumbling away into mummy dust while bare wire hits the circuit, sending pungent electrical smoke up your nose and causing you to sneeze your Life Saver into the aquarium? Less than graceful, still.

Typically, this keep-your-fingers-crossed tape was/is used to insulate wire joints that have been twisted or soldered together. But thanks to the wonders of science (and, well, shrink wrap), we have a great replacement.

Heat-Shrink Tubing

Heat-shrink tubing comes in long lengths (sometimes even spools with hundreds of feet wrapped around) and is cut to size to use. Before two wires are joined, a short piece of the tubing is slipped over one wire or the other. After the wires are soldered together, the piece of tubing is slipped over the wire junction (see Figure 6-29).

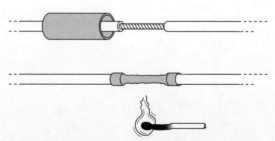

FIGURE 6-29: Slip section of heat-shrink tubing over wire (top), cover soldered connection, and heat the tubing with a lighter or flame to shrink it over the connection.

Using a lighter, heat gun, or even the shaft of your soldering iron, you then heat the tubing. The tubing shrinks down over the joint, creating a tough electrical insulation. Having a good selection of heat-shrink in various small sizes will come in very handy. Pre-snipping into ³⁄₄" sections isn't a bad idea, either.

The Well-Behaved Workspace

Hugo, the hurricane of 1989, was still packing winds strong enough to throw a car off Michigan's Mackinac Bridge when it hit the straits of Mackinaw, more than 500 miles inland. My camp, a little farther north, where I had a couple of tents set up deep in the woods, was about to be destroyed. I was in my music tent, the larger of the two, with old friend and experimental musician Marc Sloan. This double-wall tent is my outdoor workshop, jam area, and even wild mushroom dryer. The latter, drying mushrooms for preservation, is accomplished nicely because the tent roof is cut to accommodate a wood stove!

The fire in the cast iron stove was well banked with wood for the cold night coming on. We'd hiked to this tent, pitched way back in the forest and away from the main camp, much earlier in the day to sit and explore a newly bent sampler. I often do this with a new instrument: sit with a notebook and chart playing methods and switching sequences arrived at to achieve different sounds. Night had fallen by the time the remnants of Hugo hit, and by then the tent was lit only by candles.

The air molecules within a tent move in unison. If several candles are lit, their flames will dance in concordance, each replicating nearly exactly the other's movements. In Hugo's paced and rising winds there came waves of disturbance presenting as ground-level gusts, confused in direction and gaining in fury.

Through this tumultuous prelude, the flames of the candles swayed like fire dancers in unison with each other as the tent breathed. They would pause momentarily and then snap to the opposite direction as the canvas was thrown back and forth in the wind. The yellow illumination within the tent shone and fluttered as the flames cavorted in my frail shelter, painting the energies of the storm.

With the sound of a jumbo jet shearing through the trees, Hugo's gust front finally broke the forest. When the leading wind hit my small pine grove, all the tent sides simultaneously jerked inward, all three candles snapped out, and from the wood stove's mouth a wicked breath of fire blew orange embers and smoky ashes across the dark, carpeted floor. In the tent's darkness this fiery exhalation measured down-gusts throughout the night, often accompanied by the sound of snapping limbs outside the tent in all directions.

Waiting for the storm to pass, unable to keep candles lit, we settled back into this unsettling scene and, by the furious glow of the wood stove, continued to explore the circuit-bent sampler while the forest of Hiawatha was torn apart all around us, flattened by invisible giants tumbling through the trees.

Great experience. Exciting workshop. But we can do better.

Light It Up!

If drama were the object here, my fire-breathing stove would get high marks. But brushing burning coals off your notes gets tiring fast. So, let's rethink things.

Everything you do in your workshop will be easier to accomplish and more finely executed if your work area is *very* well lit. Let's consider a few options.

Fluorescent vs. Tungsten

Fluorescent lights have their fans. I am not one of them. Yes, fluorescent is nicely diffused. Yes, fluorescent is cool burning. Yes, fluorescent is cheaper to operate. Still, my experience favors tungsten's brilliance, simplicity (no ballast, fragile tubes, or possible flickering/color shift with age), small size, and noninterference with audio circuits (I've traced several fuzzy buzzes to fluorescent fixtures).

On the bright side of fluorescent fixtures, there's always the magnifying jeweler's light wherein a giant magnification lens is surrounded by a circular fluorescent tube, all mounted on a flexible neck. This classic close-work lamp is a winner, and worth trying if you need the magnification and don't mind the bulk of the unit on your bench (see Figure 7-1).

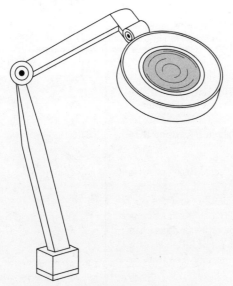

FIGURE 7-1: The magnifying jeweler's lamp
is a standard of bench work.

Multi-Spot

In my own workshop I use a combination of three tungsten lights. Above my head I use a 150-watt spotlight and a 100-watt frosted bulb, the spotlight trained on the bench from a little off to the side. But right next to the work area I have a small 55-watt quartz-halogen spotlight. This is mounted in a swivel tube, allowing it to be aimed directly at the target circuit (see Figure 7-2).

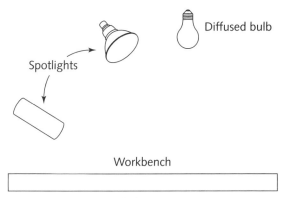

Diffused bulb

Spotlights

Workbench

FIGURE 7-2: Three-point lighting

A side light is nice because it will cut shadows, illuminating areas that the usual overhead lighting schemes might keep in the dark.

Hard shadows are minimized within this three-point lighting scheme. The proximity of the halogen lamp provides brilliant illumination. All bulbs are common and easy to replace and store.

Any way you go here, try to avoid glare while yet attaining brilliant illumination without hard shadows (hard shadows will occur if using, say, a single spot off to the side of the workpiece).

Ventilation

Because you'll be soldering and painting, ventilation becomes an important issue and something to think about. This is not to say that you need a clean room environment. But some simple considerations will go a long way.

Fans and Circulation

Soldering is often done quite casually, in assorted "well-ventilated" areas. Example? Well, your bedroom, for the most common, yet bad, example. No, the air duct in the room does not constitute "well ventilated," unfortunately. If there's no outlet for the incoming air in the form of a cold air return vent or open window you're just tumbling smoky air around and around and around.

What you want is a cross-draft across the work area, and then that air exchanged for fresh air in a constant cycle. Not so hard to do. The easiest way to achieve this is using a window fan. Just fit a fan to your window style and light a stick of incense. Walk around your workspace and observe at what spot the stream of smoke is best diverted away from you. That spot is the best spot for your soldering station as long as you position yourself upwind in the air current flowing over the workspace (see Figure 7-3).

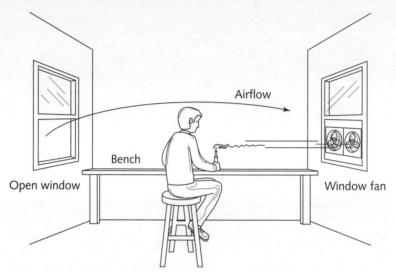

FIGURE 7-3: Creating a draft away from your soldering station

The window fan will create a draft in the room even when the air conditioning system is off, and even if the window is closed, because air will still flow through the ducts and into the room. An additional table-top fan can be used in conjunction with the window fan, positioned to blow directly over the soldering station, further encouraging airflow toward the window.

I've developed the habit of holding my breath when I solder. Inhale; solder; let smoke drift; exhale. Why not?

Building a Down-Draft Soldering Station

If you've seen indoor built-in grills you're familiar with down-draft technology. Here, even the heavy smoke of grill-top cooking is drawn downward and out of the kitchen via a blower, not unlike the blower in a kitchen range hood. Building your own is not too difficult a project.

Variants will abound here as to space and available parts, but we can generalize and allow your own design to develop to fit your workshop (see Figure 7-4).

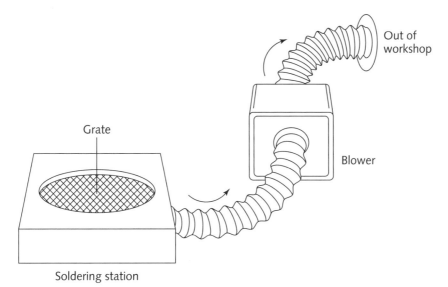

FIGURE 7-4: **Design of the DIY down-draft soldering station**

1. Find or design a shallow box and cut a 12" hole in the top and a smaller hole in the side, a hole to accommodate a blower hose.

2. Close the larger hole with a firm metal grate whose holes are small enough to keep tiny parts from falling through. (A good hardware store will have on hand a variety of perforated sheet metals in brass, aluminum, and steel.)

3. Attach a hose to the smaller hole, the hose to the suction side of a blower (blowers are often available cheap in surplus stores), and another hose from the output side of the blower long enough to go out your window.

For variations, try building a version into your workbench, able to be covered when not in use. Ideally, the blower will be situated in a far corner of the workshop, or even outside in a weatherproof enclosure. Why listen to it whirring away if you don't need to?

Now the solder smoke will try to rise but will quickly be drawn down and away, making other safeguards needless. As to painting, hold on—I cover that in Chapter 12.

Tools: At Hand But Not in the Way!

Hand tools are both the first thing you'll need as well as the first thing that will get in the way. This can be a real problem because a clean and open workspace is mandatory. No matter what storage scheme you use, follow the cardinal rule of good workmanship: As soon as the tool is no longer needed, quickly put it away in the place you've dedicated to it.

Building a Way-Cool Tool Station

Hanging tools on traditional perf board works, but there's a lot of reaching involved. Solution? The lazy Susan tool caddie. Although I don't know this particular Susan, could be she was just preoccupied. What we know for sure is that what goes around comes around, never better demonstrated than by this particular contraption.

If you don't already know, a lazy Susan is a round platform able to rotate on its base. The bearing alone can be purchased for constructing your own if a lazy Susan can't be found (bearings in hardware stores; Susans in cooking and kitchen supplies).

When you have a wooden lazy Susan to convert, simply drill holes for the most-used tools to slip down into. Be careful not to drill all the way through the top of the turntable or the tools will penetrate to the bottom of the turntable and jam the rotation.

Include screwdrivers, socket drivers, reamers, and any other tools that will locate in a drilled hole. Also consider including a hole or two for inserting wooden dowels to serve as rods on which to stack spools of wires and solder (see Figure 7-5).

FIGURE 7-5: Revolving tool caddie

Parts Containers, Conventional and Non

Nothing wrong with the usual multidrawered parts cabinets available in hardware and electronics stores. If you go this route, be sure to buy sturdy cabinets. Although lightweight when empty, the filled cabinets will be quite heavy. Thick walls and drawers will pay off here. But there are other ways to go.

Aren't Jewelry Boxes for Jewelry?

My favorite nonconventional parts container is a certain kind of jewelry box that I seem to routinely find at the secondhand shops I frequent to look for bendable toys. These boxes use a scissors-like mechanism to not only automatically open a front drawer when the lid is lifted, but also often to unfold symmetrical inner tiers, rising like petals on an opening blossom. All inner surfaces are divided into small containers, perfect for many electronic components. Closed, they are dustproof. They're quite cheap, at least locally, usually selling for two or three dollars each (see Figure 7-6).

FIGURE 7-6: Mechanical jewelry box for part storage—glass eyes, in this example

Skippy vs. Organic Almond Butter Jars

Your granddad saved his glass jars to keep nails, tacks, and screws in. Grandpa was right. And recycling is as smart today as it was back then. Okay, smarter.

Glass jars with screw-on lids are great for electronic parts. Because they're dust-proof, see-through, and available in all sizes, storing components in these slow liquid containers (yep, glass is a liquid flowing so slowly it seems solid) is a good, free choice. As to regular versus organic, choose organic. The glass isn't any better, but you'll be.

Can o' Pens

While we're looking backward to the old-fashioned workbench for solutions, don't pitch that empty can of beans. You'll be using pens frequently to take notes as well as to mark circuits, and you'll need a place to keep them. The bean can holds a certain attraction to many "pen pals" as well. In no time you'll find your retractable X-ACTO knife, heat-shrink tubing, and mechanical compass all vying for this handy column of space. In fact, you'll soon be serving eviction notices.

Parts Magnet

Everyone has a blown speaker lying around. If you can salvage the magnet, it will serve you well as a small metal parts holder, perfect for all the screws you'll be removing from instrument cases as you disassemble things. Mine is close at hand, firmly gripping the base of my scrolling saw.

Choosing Your Work Surface

You and your workspace will become very intimate. You'll draw on it, read on it, solder on it, and more. You'll need it to cushion as well as to provide firm backing. You'll want it to be convertible.

Hard, Soft, or Super-Soft?

All! For drawing on a sheet of paper (note taking, circuit sketching), I have a thick sheet of tempered glass at the workbench where I sit. For general circuit work I cover the glass with a terry cloth hand towel. This keeps dropped parts from rolling far, and secures circuit boards and instrument cases from slipping around while working on them. For final-stage work on painted instruments I cover the glass with a thicker, deep nap towel. This extra cushion keeps delicate finishes from marring during the final wiring steps.

Please, Have a Seat

I hope you're sitting down, because I have something shocking to tell you. The make-or-break bending session item just might be your chair! What you're looking for is a good rolling office chair with swivel, lumbar support, adjustable height, adjustable arms, and back tilt. Being comfortable will make any work session much easier.

Dr. Frankenstein, May We Proceed?

In a very real sense, you are creating new life here as you, along the lines of Dr. Frankenstein, inject electricity into your reconfigured, stitched-together circuit. The biggest difference is that while Shelley's Frankenstein story was fiction, circuit-bending is fact. Time to raise the kites into the storm.

part

III

Soldering Your Way to Nirvana

Everyone seems to be apprehensive about learning to solder. I was, too. So much that I grabbed that tube of "Liquid Solder" from the hardware store back in 1968 and, with high hopes, took it directly home.

Not that I didn't have a soldering gun—I did. I'd won it in a bet with my chessmate as to whether James Thurber was yet among the living. My luck being better than James's, I stashed the gun for future misuse.

Liquid Solder seemed easier than the gun, if only it lived up to its name. Opening the tube of gunk (which smelled exactly like model cement), I tried my best to join wires with it. It was silvery, after all. And it did say "Solder" on the label, loud and clear.

All I succeeded in doing, in the end, was to ruin a bunch of good components by smearing the nonconductive silver model cement all over the solderable leads, insulating them and making them nonsolderable forever. Well, my world and welcome to it!

Solder Is Your Best Friend!

If you're already solder-savvy and don't need a refresher, feel free to skip right over this chapter to the next (Chapter 9). If you want to take a look at a couple of bending-centric soldering techniques and tricks, you might want to skim this chapter for anything new. And if you've never soldered anything before, this chapter's for you.

One skill *critical* to circuit-bending success is good soldering. Many people are scared of soldering. Don't be. It really is easy. You can even look at it as a hot-melt glue that conducts electricity because, well, that's what it is.

How Solder Works

Solder is a combination of metals that have a low melting point. And, of course, the metals conduct electricity. The rosin in the solder's core serves as a "flux," melting along with the solder and making the metals to be joined more receptive to the solder itself.

in this chapter

- ☑ Solder is your best friend
- ☑ How solder works
- ☑ Everyday bent soldering needs
- ☑ Special soldering tricks

Solder, when heated and in liquid form, has properties you've seen often in water and other fluids. For example, just as water flows and is absorbed into crevices, so is liquid solder. The trick with solder is that the surfaces involved must be metal, must be clean, and must be *hot*. At least hot enough to melt the solder.

Speaking of hot, now's the moment to reinforce the fact that soldering involves temperatures at the tip of the iron hot enough to give you a nasty burn (500 to 800 degrees, typically). If you handle the soldering pencil carelessly, as in you leave it up against your gunpowder collection or on your original parchment document proving that the Chinese discovered America before the Europeans did, you'll be very, very sorry.

Always return the hot iron to its rest, and be sure that it's fully inserted, or the tip can heat the rest, which in turn can melt the handle of the iron. And *never* touch the hot tip of the iron, even just to see whether I'm kidding.

Practice Makes Perfect

Never is the statement "Practice makes perfect" more true than with soldering. With a little practice, you'll come to know the steamy personality of solder and soon find yourself using its subtleties to your advantage. I cover some unusual uses in a moment, but here are the basics.

- Use a 25–40 watt soldering pencil, as described in the tools section.

- Keep the tip wiped clean with a wet sponge or cloth.

- Keep the cleaned tip coated (tinned) with a little melted solder.

- Be sure that metals to be joined are bare and clean.

- Sand the traces on circuit boards to bare metal before soldering (see "Soldering to the Circuit Board," later in this chapter).

- Make sure that both metals to be joined are hot before the solder is fed into the junction. Be sure to heat the junction hot enough to easily melt the solder—a good joint looks almost chrome-like and very smooth; a bad or "cold" joint looks dull, is often "balled up," and can even be fragile.

- Be sure to keep wires *absolutely still* for a few seconds as the joint cools. Another sure-fire way to end up with a cold joint is to move the metals being connected before the solder hardens.

- Always place the flat side of the "screwdriver" or "chisel" tip against the metal you're heating, because doing so gives greater surface area contact.

- Use only enough solder to smoothly coat the junction and flow fully into the connection. Avoid using too much solder—big solder blobs are to be avoided.

A Beginning: Splicing Wires

If you're already solder savvy, great! Feel free to skip on ahead. If not, here are a couple of exercises to get you started.

1. Using your wire stripper, strip about ¾" of the insulation off the ends of a couple of short lengths of wire-wrap wire (see Figure 8-1).

FIGURE 8-1: Stripped 25-30 gauge wire-wrap wires

2. Twist the bare ends together, as illustrated in Figure 8-2.

FIGURE 8-2: Twisted wire-wrap wires

3. Place the tip of your hot soldering pencil against the bare wire junction, about in the middle (see Figure 8-3).

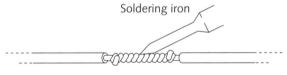

Soldering iron

FIGURE 8-3: Heating the wires

4. After two to three seconds, touch the end of the solder strand to the hot junction (see Figure 8-4).

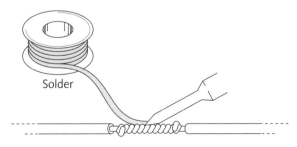

Solder

FIGURE 8-4: Applying solder

5. Keeping the soldering tip in place, feed the solder into the junction as it melts until the twisted wires are coated with solder (see Figure 8-5).

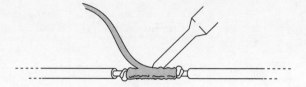

FIGURE 8-5: Solder melting into hot wires

6. Remove the solder and then the pencil tip; allow the junction to cool for a few seconds before moving.

Remember, good connections use solder modestly, not excessively (see Figure 8-6).

Good joint   Overkill

FIGURE 8-6: Solder should coat junction without much excess.

Keep practicing this connection with new sections of wire until you feel confident. When you have this down, you're over the hump.

In this example, the soldering time is quick because there's not very much metal to bring up to temperature. Be aware though that in many soldering applications, you'll have to warm much more metal than the thin wire-wrap wire you're practicing on in the preceding example. Still, the technique is the same.

Everyday Bent Soldering Needs

As you build the projects in this book and continue on your own, you'll employ all the common soldering techniques. Here are the most important ones to know.

What's "Tinning"?

A term you'll see often in electronics, tinning simply means to thinly coat with solder. Melting a little solder onto the tip of your soldering pencil after cleaning it is tinning it, a process that helps keep the tip in good condition and solder-ready.

More often, tinning is done to metals to be joined prior to actual soldering. This tinning makes the connection quicker, easier, and more electronically secure. Try the following if you're not already familiar with tinning.

1. Strip about ¾" of the insulation off the end of a multistrand wire (see Figure 8-7).

FIGURE 8-7: Stripped multistrand wire

2. Twist the bare wire strands tightly together by holding them still between thumb and forefinger and spinning the wire where the cut insulation begins with your other hand (see Figure 8-8).

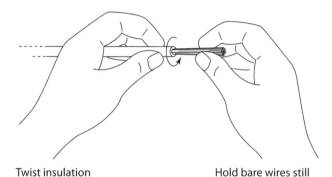

Twist insulation   Hold bare wires still

FIGURE 8-8: Twist insulation while holding stripped wires.

3. Hold the hot soldering tip against the bare twisted wire (see Figure 8-9).

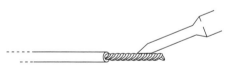

FIGURE 8-9: Heating the twisted wires

4. After a few seconds, feed solder into the tip/wire junction until melted solder flows into the twisted bare wires (see Figure 8-10).

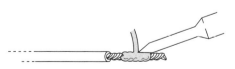

FIGURE 8-10: Solder flowing into twisted wires

5. Remove solder and then tip; allow tinned wire to cool (see Figure 8-11).

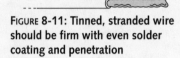

FIGURE 8-11: Tinned, stranded wire should be firm with even solder coating and penetration

6. Inspect tinned wire to confirm that solder has flowed evenly into the wire, providing a uniform coating.

The loose wires have now been joined together to prevent fraying and straying against other parts of the circuit next to where the wire is meant to be soldered. Important: *Always* tin *stranded* wire before soldering it to whatever.

Soldering to Switches and Pots

I already covered the wiring of switches and pots back in Chapter 6. The actual soldering is simple.

Both these components have soldering lugs ready for soldering to; some are even pre-tinned. Just run your stripped (and pre-tinned, in the case of stranded) wire through the hole in the lug and crimp it to hold it firmly in place (see Figure 8-12).

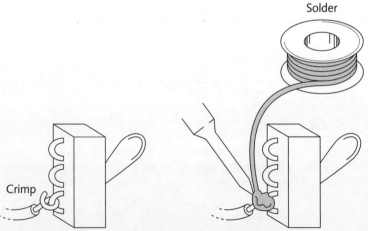

FIGURE 8-12: First crimp wire, then solder. Clip away any excess wire.

Solder this junction just as you did the wire connection in the first exercise.

Soldering to LEDs

Soldering wires to LEDs is just about the same as soldering wires to wires. The two leads of an LED are usually pre-tinned. Just wrap the stripped wire around the LED lead a few times and solder as usual. Here, though, be sure to solder as quickly as possible, because an LED is a heat-sensitive component and will fry if it gets too hot. In fact, the actual lead construction of LEDs is metal heavy in an attempt to absorb heat away from the actual light-emitting element.

Because we'll always be using thin wire to solder to LEDs, wire that heats quickly, heat build-up shouldn't be a problem. Soldering to the far ends of the LED's leads, away from the LED housing itself, will keep heat down. If you still want a little added protection, there's always the heat sink.

Heat Sinks

LEDs are not the only heat-sensitive components you'll be dealing with. Transistors, present in many circuits you'll be bending, are ready to fry as well. That's what clamp-style heat sinks are for.

It's not a bad idea to use a heat sink on transistor leads if you find a need to solder to one. If so, just clamp the heat sink to the lead between the body of the transistor and the soldering point and solder as quickly as possible, the same as if you were soldering to an LED (see Figure 8-13).

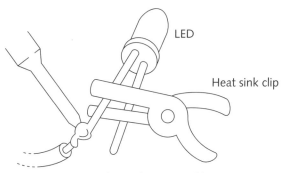

FIGURE 8-13: Heat sink goes between soldering iron tip and component.

The same position would go for regular diodes, which are also heat sensitive. Heat sink clamps always go between component body and the part of the lead you're soldering.

Soldering to Speakers

Often, you'll be deriving line outputs from speakers already built into the devices you're bending. This is accomplished by soldering wires to the speaker's two terminals (soldering lugs, again). The only new thing here is that the terminals will already have wires soldered to them. You simply need to take care that in soldering you don't de-solder any wires already there.

Soldering to the Circuit Board

At times you will need to solder wires directly to the thin metal traces on a printed circuit board. These intricate metal traces that connect components on the board are *delicate*. They're probably also coated with a transparent coating of some kind that must be removed before soldering to the trace is even possible. A careful approach is needed.

If you're good with your Dremel drill, you can remove the clear (sometimes colored) overcoating from the trace with a small burr bit running at medium speed. Too much pressure will cut through the trace! Be extremely cautious here. The burr bit may want to run across the board if not controlled well. Carefully and *lightly*, stroke the trace with the spinning burr just until bright metal is seen. Expose about ¼" of the trace in this way (see Figure 8-14).

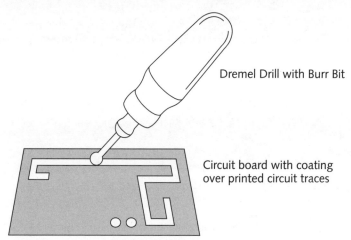

Dremel Drill with Burr Bit

Circuit board with coating over printed circuit traces

FIGURE 8-14: Use only the lightest pressure to remove a printed circuit's top coating.

The coating can also be removed by carefully scraping with a small blade, such as that of the chisel-tipped X-ACTO. Even if the trace doesn't seem to be coated, use fine sandpaper or steel wool to remove any top surface, including tarnish, before you attempt to solder.

After the trace is cleaned to shiny bright metal, hold the tip of your soldering pencil against it. Be sure to clean the tip first (very important—the more delicate the work, the more critical the need for a super-clean tip).

Apply a tiny bit of solder to the tip/trace junction, just enough to leave a small bulge of solder on the trace when you remove the soldering pencil tip (see Figure 8-15).

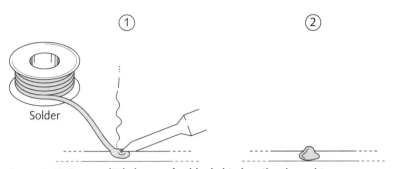

FIGURE 8-15: Leave a little bump of solder behind on the cleaned trace.

That's it! The little blob of solder will now serve as your soldering point. You'll need to heat it only a moment to be able to solder your 25–30 gauge wire-wrap wire to it (see Figure 8-16).

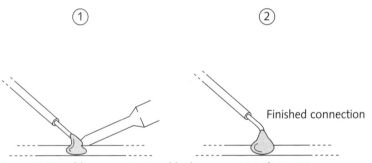

FIGURE 8-16: Solder your wire to solder bump on printed circuit trace.

Do remember that the actual metal of the printed circuit trace is fragile. If you leave too large a blob of solder on it and strike the solder accidentally, you can break the trace right off the board. Likewise, even with a small amount of solder on the trace, an accidental tug on the wire you soldered to it can tear the trace right from the board. Always treat these connections with care.

Soldering to Component Leads

As discussed in the preceding LED section, component leads are easy to solder to. Simply wrap your wire around the component's lead and solder as quickly as possible. A medium-sized jeweler's screwdriver will help you manipulate wires in tight places (such as on the short leads of an electrolytic capacitor already soldered to the circuit board) in case you have trouble wrapping the couple of needed turns by fingertips alone.

In the instance of heat-sensitive components, do use a small heat-sink clip, as described previously.

Follow That Trace!

This is cool. Say that you need to solder a wire to the lead of a component mounted so closely to the board that you can't get to the actual lead—not enough room for the soldering tip to get in. All you need to do is to visually follow the printed circuit trace and solder to the trace where it becomes better available elsewhere on the board. Even if you have to follow the trace to the far side of the board, soldering to the trace will be the same electronically as soldering to the hard-to-get component lead.

Let's make it easier. Say that as you follow the trace you see that it's connected to other component leads here and there. Yes, as you just guessed, soldering to any of these leads is the same as soldering to the hard-to-get lead. Now you don't have to scrape a circuit trace at all (see Figure 8-17).

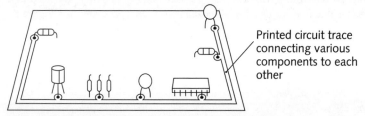

Printed circuit trace connecting various components to each other

FIGURE 8-17: No matter where or to what component lead along a single printed circuit trace you solder a wire, you get the same electrical connection.

Soldering to IC Pins

Often you'll find great responses when IC pins are brought into the new wiring scheme. But soldering to IC pins is tricky at first. Don't worry—it will become routine.

First, follow the printed circuit trace that the IC pin runs into and see whether there's an easier place to solder to. As detailed previously, soldering to the actual trace to which the pin goes is the same as soldering to the IC pin itself. If you can't find a better place along the trace to solder to, then the pin it will be!

The trick here is not only to pre-tin the IC pin but also to leave a tiny blob of solder behind, just as I explained for soldering to the printed circuit trace. The operation is very fast because an IC pin is small and thin, contains very little metal, and therefore heats quickly. A quick connection is needed when soldering to ICs to help minimize overheating the IC's internal circuitry.

1. Tin your tip with a little solder.

2. Place the chisel side of the tip of the soldering pencil against the leg of the pin and almost immediately apply the solder to the pin exactly where the soldering tip meets the pin (see Figure 8-18).

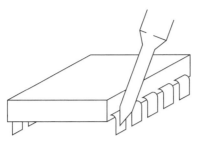

FIGURE 8-18: Heat the IC pin for just a moment before feeding solder to the tip/pin junction.

3. Remove the solder and tip just a moment after the solder begins to melt onto the IC pin, leaving a tiny bulge of solder behind (see Figure 8-19).

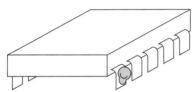

FIGURE 8-19: A small solder bump left behind on the IC pin

With a sufficiently heated tip, this entire procedure should take no more than a second.

4. Just as in the example of soldering to the cleaned printed circuit trace, the solder blob is heated and the stripped end of your wire-wrap wire is fed into the hot mass (see Figure 8-20).

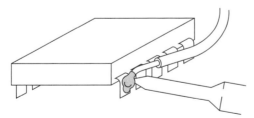

FIGURE 8-20: Feeding stripped wire wrap wire into reheated solder bump

5. Remove the tip and hold the wire absolutely still, as in all soldering, until the solder hardens (a second or two).

Another approach to IC pin soldering involves molten solder transport, discussed in a moment (see Figure 8-25)

It's also sometimes possible to solder to the IC pins emerging on the bottom of the board (see "Playing Hooky," which follows).

Special Soldering Tricks

Everyone develops his or her own techniques in tool using. Some of these techniques break cardinal rules, including one I'm about to teach you. Still, these little tricks can take you a long way.

The Stationary Iron

Rather than bring the soldering tip to your work, you might sometimes want to bring the work to the tip. This makes jobs such as tinning a bunch of wires a breeze.

Simply lay the hot iron on your workbench, tip pointing right at you. Now bring the solder and wires to be tinned to the hot tip. Touch the solder to the tip and feed the wire into the solder as it melts, right in the solder drop that forms under the tip (see Figure 8-21).

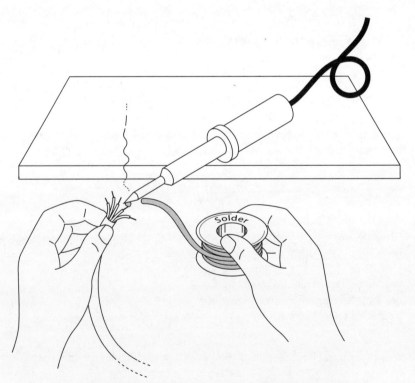

FIGURE 8-21: Using the stationary iron for tinning a series of wire ends

This will also streamline joining wires to each other, soldering wires to hard-to-position parts, and many other applications. After all, the soldering pencil is often the heavier and harder item to position within the soldering triad (iron, solder, and component). A corded tool only adds to the potential trouble. Keep this in mind—opportunities will arise.

Playing Hooky

Keeping wires where you want them while soldering can be a problem. Sometimes there's little room to wrap a wire around a soldering point (as in a mini resistor mounted flush with the board), or the point itself is just too small (such as an IC pin emerging from the back of a printed circuit board). Preshaping the end of the wire may be the solution.

Form the end of the stripped wire into a tiny hook that can be slipped over the target IC pin or component lead (see Figure 8-22).

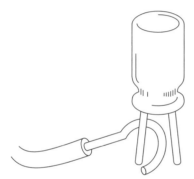

Figure 8-22: Make a hook on the end of the wire for locating it on difficult areas.

Slip it over, solder, and finally clip away any extra wire remaining.

Wrap That Wire!

Even with all the aforementioned techniques, sometimes the wire you're trying to solder just won't stay in place. If the wire is long enough, after hooking the stripped end over the target connection point you can wrap the wire's other end around a nearby component on the board. This will hold the wire in place for soldering (see Figure 8-23).

If you have no handy component for wrapping, you can always use a piece of masking tape to hold the wire still.

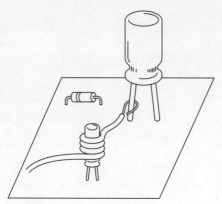

FIGURE 8-23: Use a nearby component
to keep the hooked wire in place.

Molten Solder Transport (Don't Tell)

In traditional soldering, the idea of transporting melted solder around on the tip of a hot iron is unthinkable. The fear is that the solder will drop off the iron, falling onto the circuit and sending you directly to the repairs section that follows. It's a real concern, too. However, if used with care, molten solder transport is just fine.

Let's say that you're soldering a wire to an IC pin, just as described previously. But you fear that the solder you left on the pin in the tinning process wasn't built up enough to solder the wire when melted. You need a little more solder. Using the heat of the tip, you can "cut" a small section of solder, about $1/16$" long, off the roll. This $1/16$" section of solder will immediately draw itself onto the hot tip and remain there (see Figure 8-24).

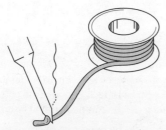

FIGURE 8-24: Cutting solder off
the roll with your soldering tip

If you hold the wire against the IC pin and then bring the molten bit of solder to the wire/pin junction, the solder on the tip should flow onto the IC pin, soldering the wire nicely in place (see Figure 8-25).

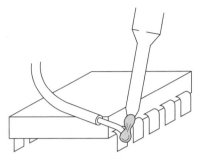

FIGURE 8-25: The small bit of transported solder will flow onto the IC pin.

I do this all the time. Even all by itself (without your pre-tinning the IC pin first), this molten solder transport works well for this type of small-area connection.

Keep in mind that the rosin in the core of the solder boils away as the solder is heated; you can see this smoky event if you watch. Because the rosin serves to help hot solder adhere to the metals being soldered together, the best connections are made while the rosin is still boiling. In molten solder transport, the rosin might boil away before the solder makes it to the connection.

The only answer here is to be quick in getting the freshly cut hot solder to the connection. Yes, this increases the chance of dropping hot solder on the board. Still, with practice, traveling molten solder is a legitimate soldering technique.

The key caution here is to never carry more solder on the tip than is needed for a connection. In *any* solder connection, think small over large: Use as little solder as needed to smoothly coat the joint, and no more.

Fixing the Inevitable Soldering Mistakes

So, you trusted my advice, tried the molten solder transport trick, and now have a glob of dropped solder on the circuit board. All is not lost.

Usually in the case of dropped molten solder, repairs are not too difficult, because the solder has fallen on cold surfaces upon which it will not fully adhere. Careful scraping and prying with an X-ACTO blade will usually remove these thin splobs (a splob is a splattered blob).

Two seat-of-the-pants methods exist for solder removal in the instance of too much solder having been applied to a single connection, or solder straying into easy-access areas where it doesn't belong. The first method is to see whether it will adhere to the cleaned tip of your hot soldering pencil. If so, remove the solder by drawing it up on the hot tip, cleaning the tip, and repeating until the solder is removed (or until the residual solder can be scraped away).

The second technique can be touchy to disastrous, depending on your standing with the gods of fortunate soldering. Heat the solder to liquid state and quickly, while the solder is still hot, sharply tap the board against the workbench in such a direction as to fling the solder off and away from any other components on the board. Don't try this with circuits where the molten solder might run under ICs, into IC pins, or any other place that might complicate rather than solve the problem. Reserve it for simple circuits. Pretty radical solution, but it has worked for me many, many times.

For solder mistakes involving more difficult situations, such as solder accidentally flowing between two adjacent pins on an IC, specialized tools are made for this kind of touchy solder removal. The best of these is a simple "solder wick."

Made of braided wire, the wick will absorb molten solder the way a sponge absorbs water. In use, the wick is placed against the unwanted solder and then heated with the soldering iron tip (see Figure 8-26).

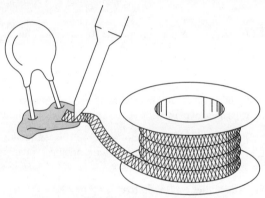

FIGURE 8-26: Heat the de-soldering braid against
the solder overflow.

You'll see the solder melt into the braid, with any luck leaving only a small bit behind, easily scraped away. These braids can be trimmed to shape to fit in tight spaces, and after use you simply clip the solder-filled section away.

There are also small solder vacuum pumps that draw melted solder up and into their reservoir. I prefer the braid, but you might try both and see which system you feel best with.

Before you begin to work on the project section of this book, it would be a good idea for you to practice all the soldering techniques I've discussed. Splice wires, tin multistrand wire, solder to switches and pots, LEDs, speakers, circuit board (component leads such as resistors and capacitors as well as to printed circuit traces), and IC pins. Try the stationary iron, molten solder transport, and the solder removal techniques as well.

For this practice session, you'll need to use some of the components you've collected (pots, LEDs, and switches) as well as have a circuit to work on. Any sound toy with the batteries removed will probably do, as long as it has a circuit board and speaker. At secondhand shops, such a toy will cost only a couple dollars. Try soldering wires to the circuit board in as many ways as possible and until you're comfortable with the procedures. Better yet, if you have a dead circuit of any type lying around, try soldering different kinds of wire to it, anywhere you can (see Figure 8-27).

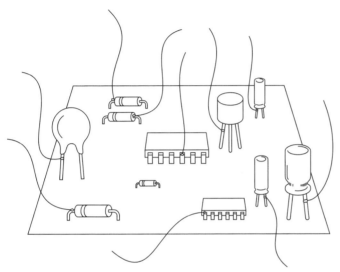

FIGURE 8-27: Practice soldering to everything on a dead circuit.

Designing Your Personal Alien Orchestra

"Specimen quality" and "museum quality" are familiar terms to amateur mineralogists such as myself. On a trip through Canada I visited an amethyst mine in Thunder Bay to search through their "tailings," the rough stone already gone over in search of fine grade amethyst. The idea was that specimen-quality stones might still be amidst the rubble and were for sale by weight to persons lucky enough to find them.

Within a few minutes I spotted a sparkly cluster of jagged-edged stones, looking as though they were all stuck together. Turning this clump over, I was thrilled to discover that this was the base of a single large crystal, all facets of the natural structure unmarred and terminating into a sharp point. A keeper.

Back in the office, having taken the crystal in to be weighed, I was immediately interrogated: Where exactly did I find the crystal? Turns out that the iron inclusions in the amethyst crystal I found made the specimen very rare. Museum quality, in fact. And 50 cents by weight.

The circuit-bending process is just like this—like hunting in a gem mine. Discarded circuits are very much the tailings of our consumer electronics market. And like the tailings of the amethyst mine, they contain museum quality gems, "by weight," for anyone prepared to search.

Finding Great Circuits for Bending

Okay! You're confident with your soldering, you've put salve on your burns, and you're itchin' to bend. You hit your local Salvation Army retail outlet and the avalanche of abandoned toys falls upon you. As you regain consciousness, an electronic voice says, "Wrong. Try again!"

Fabulous. You've just found a Touch & Tell series talking game with the batteries already installed. But this kind of luck is rare. More often you'll see in the mass of toys a bewildering spread of potential bendables, without batteries and silent. You could buy them all and get the kid next to you really mad. Or you could be a good scout.

Be Prepared

To be prepared as a bend scout is as simple as having a little money and a lot of batteries: Money for the secondhand sound toys; batteries to see whether they work. You'll also need two small screwdrivers to open the battery compartments: a regular and a Phillips head. I have a key chain device that has both drivers as well as a knife blade and wire cutter, awl, and rasp (see Figure 9-1). What's nice about this is that it's no larger than a key and is always with me. Keep your eye out for these mini tools.

FIGURE 9-1: Multitool including standard and Phillips drivers

As for the batteries (which you'll collect a lot of, finding them in toys you'll buy), carry with you six of each of these sizes: C, AA, and AAA. To be sure they're fresh, get a stand-alone (not battery-powered) battery checker. Carry a nine-volt battery, too.

Some sound toys use D batteries as well, but rarely. If you want to add six of these to your backpack, you're quite the soldier. Bend all that you can bend!

What Can a Voice Tell You?

Until you open a toy's case to see how accessible the actual circuit is, all you have to go on are the hints provided by the voice and the control array. But these can really tell you a lot.

Controls and Voice Selection

Generally, the more control and voice options, the better. A musical keyboard or other toy with lots of different sounds is likely to have more circuitry to explore than are simpler devices with fewer front-panel choices.

Beyond this is the actual quality of the voice. Is it kinda fuzzy? Or is it high fidelity? A strong, hi-fi original voice often translates into a strong, hi-fi bent voice. Take a good listen.

Your Most Important Tool: The Bending Probe

The most important tool for circuit-bending can be as simple as a short section of wire. After all, you're just going to be using a wire to connect one circuit point to another to see what happens. Trouble is, you'll need something easier to handle than a wire. You need pointy ends on the wire to probe into tight places. And you'll need firm handles to hold on to. You have choices.

Buy or Make?

Electronic meters, as with volt/ohm meters, come with a set of test leads. One end of each lead has a plug on the end to plug into the meter; the other end of each lead terminates in a thin probe housed in a plastic handle. Occasionally you'll find probe tips available by themselves. Just solder a probe tip to each end of a 12" section of insulated wire and you're in business.

The Mutilated Test Lead Approach

If you prefer less work and don't mind spending a little more money, just buy a set of test leads by themselves and transform them into your bending wire. Simply cut the plugs off and solder the two wires with probes at the ends to each other. Here's the procedure.

1. Cut each wire about 6–10" from the probe's handle (see Figure 9-2).

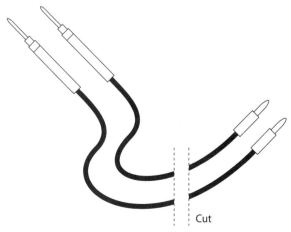

Cut

FIGURE 9-2: Snip the small plugs off the test probes—you don't need them.

2. Strip about ¾" of the insulation off of each end (see Figure 9-3).

FIGURE 9-3: **Strip ends of probe wires.**

3. Slip a 1" section of heat-shrink tubing over one of the wires (see Figure 9-4).

Heat-shrink tubing

FIGURE 9-4: **Thread heat-shrink tubing over one of the wires.**

4. Twist the wires together and then solder, allowing solder to flow well into the connection—wire must be hot; be patient (see Figure 9-5).

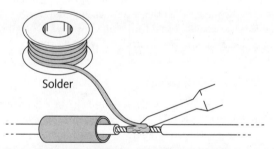

Solder

FIGURE 9-5: **Twist bare wires together and solder.**

Note | Tinning the wires previous to soldering here is an option, but the wire-to-wire junction is pretty good, and tinning can be bypassed in this in-line splice as long as you achieve thorough solder penetration.

5. When the solder connection is completely cool, slip the heat-shrink tubing over the connection (see Figure 9-6).

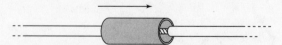

FIGURE 9-6: **Slip heat-shrink tubing over connection.**

6. Using a match or a butane lighter flame (or by stroking the tubing with the hot barrel, not tip, of your soldering iron), heat the tubing just until it shrinks down around the soldered connection (see Figure 9-7). Don't overheat it. If you do, it will burn and crack away.

FIGURE 9-7: Heat the tubing to shrink it over soldered joint.

The Classic Alligator Clip Leads: Pros and Cons

Alligator clips have metal jaws for clipping onto circuit points. An alligator clip wire (or lead) is a wire with an alligator clip at each end. If you clip a metal jeweler's screwdriver to each end of an alligator clip wire, you'll have in your hands the traditional circuit-bending wire, embracing the electronic equivalent of good, bad, and ugly (see Figure 9-8).

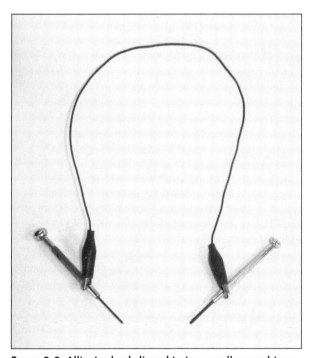

FIGURE 9-8: Alligator lead clipped to two small screwdrivers

The Good

The good is that you can buy bags of alligator-clip-terminated test leads very cheaply. And because all benders have metal jeweler's screwdrivers at hand . . . presto! The clip-the-screwdriver test lead is born.

The Bad

The bad is that the clips tend to unclip as you use the system, making for some unneeded fumbling around.

And the Ugly

The ugly is when the alligator clip that just decided to go AWOL lands itself on the most sensitive IC pin on the board, making a cute little spark and sending your Speak & Spell back to eBay as a "parts car."

True, I've bent countless circuits via the alligator clip wire. And this is still how I get new benders right into things. But do scrape together some version of the test probe version, above. Or . . .

Order the Rare, Ready-to-Go, Pin-Tip Test Leads

Though hard to find, a probe set is still manufactured whose leads are terminated in "pin" tips, perfect for bending. Part # GC 12-1647, these can be ordered from Pembleton's (e-mail: pembletonelectronics@yahoo.com).

Searching the Circuit-Board

Here comes the really fun part! The following guide walks you through any circuit you want to bend, as long as the circuit meets the requirements for bending:

- The circuit must be running on battery power of 6 volts or less.

  Why 6? Not because higher-voltage circuits aren't bendable. They are. But as voltage increases so does the risk of applying too much voltage to sensitive parts of the circuit as you jump electrons from here to there. This book reflects the majority of my experience, which, in turn, is with circuits of six volts or less. That happens to be both the voltage of the many super-productive as well best-sounding circuits you're likely to find.

- The circuit contains a speaker allowing you to hear what's up.

- The circuit is expendable.

  Expendable because circuits fry. I still fry circuits and you'll fry some, too. Remember to buy your circuits secondhand from charity outlets. It's cheap for you, helps the charity, and keeps plastic out of the landfills or incinerators. And even with burn-outs, circuit-bending is still one of the less expensive arts to explore.

Where Do You Start?

The first thing you'll need to do is expose the circuit. Using whatever screwdriver is needed, remove the screws holding the toy's halves together and slowly separate.

Don't be intimidated by the circuit. Nothing's going to explode and you're not going to get shocked. You're just going to hear, with a little luck, some very strange sounds.

First Things First: Taking Notes

As you explore the circuit you'll want to keep track of your discoveries in one way or another. Because this can get pretty complex depending on the number of bends, you might find yourself opting for one of the more involved charting procedures that, although more time consuming, not only can make charting easier but also serve as superb notes for later reference.

Different Ways to Chart Your Bends

The three main routines for keeping track of your discovered bends are as follows: on-circuit, by which you mark your new bends directly on the circuit board; separate drawing, by which you make a sketch of the circuit board and keep notes on the sketch; and image capture via scan or photo, by which you take notes on an accurate reproduction of the circuit printed out on your computer.

The Pen Is Mightier Than the Pencil

The most casual way to chart your new bends is directly on the circuit. Use a Sharpie brand permanent ink Ultra Fine Point felt-tip marker. Buy a set with as many colors as available.

When you find a good bend by, say, placing the bending wire between the lead of a resistor and the pin of an IC, you can then mark this connection so that you can solder it later. Choose a color from your pen set and put a same-colored dot at the resistor lead and the IC pin so that you know to connect them later. As you proceed, you can color code other connections you discover.

In addition to color coding, you can use one color only and just draw lines from point to point. This will work just fine on simple circuits with easy component access. But color coding is quick and clear, and the more color in life, the better.

Interesting fact here: The pen *is* a mightier tool than the pencil for on-board charting. Why? As mentioned earlier, the line drawn with a pencil can conduct electricity. So in using a pencil in point-to-point line drawing on the circuit, you run the risk of drawing actual *live circuits* as well. Cool. But not at the moment.

Drawing the Essentials

If the circuit is not too complicated, you might want to make a representational diagram of it on a handy piece of paper. In the circuit shown, all component leads connect to printed circuit traces on the back of the board; therefore all we need to chart is the back side (see Figure 9-9). In the example of boards with traces on both sides, sketch the top side, showing all the mounted components in as precise a way as needed for reference.

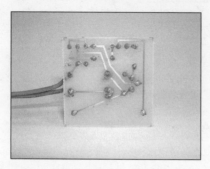 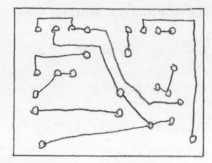

FIGURE 9-9: Redrawing the essentials of a circuit for charting

Scanning and Photographing Your Circuit

This is a great way to keep notes. Either take a digital picture or make a scan of the circuit; then, on your computer, produce a color print. With your pen set you can now keep notes on a sharp representation of the circuit itself. You'll find this a wonderful aid or future reference and will probably assemble a number of these scans in a notebook for your workshop records (see Figure 9-10).

All that said, when you go back to solder the connections you've made notes on, it's nice to see the points premarked on the board waiting for you. I often chart on the board as well as on a scanned representation at the same time. Recommended.

You: BEAsape

With all this point-to-point wiring I've been talking about, you'd suppose that's where I'd have you start, right? Nope. I'll start by turning you into a BEAsape.

If you have a photographic memory, you may add to that my green envy. For the rest of us, as mentioned in the beginning pages of this book (Chapter 1), BEAsape stands for Bio-Electronic Audiosapien. And that's the new creature you become as you merge with an audio circuit, your body now an active on-board component, conducting electricity.

Body-Contact Search

What you're trying to discover right off the bat is whether you can throw the circuit's operational parameters off and therefore change its sound by making bodily contact with it. To put it simply, you'll touch the live circuit and listen for changes.

Bare Fingers

Who wouldn't want to do such a thing barehanded? And for the initial body-contact exploration, this is fine.

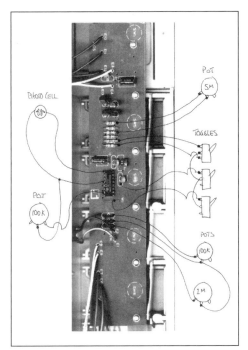

FIGURE 9-10: Using a scan of the circuit for charting

Here's what you do.

1. Turn the toy on and have it make a noise.

On keyboards there's usually a demo tune you can get to repeat over and over until someone nearby gets irate and forces you to explain "art." Other times, you can tape down a key or button to keep a sound going.

2. With the instrument making a sound, begin touching the various circuit traces and components with one of your fingertips, listening for any change in sound (see Figure 9-11).

3. Using your charting method of choice, mark any spots that changed the sound when touched.

If you were lucky enough to find areas that altered the sound when touched (usually in pitch or timbre), you'll have a baby-simple task of soldering wires to these points later on so that they can be connected to metal body-contacts. I mark "BC" next to body-contact points on the board.

Don't ignore a point that just affects volume when touched. This will allow "tremolo" when touched rhythmically. All body-contact changes can be used to animate a voice, no matter how extreme or subtle.

FIGURE 9-11: Touching the circuit while listening for changes in sound

There are times when (maybe only) touching *two* points simultaneously will initiate a body-contact response. To look for this possibility press a finger of one hand, the "stationary" hand, to a circuit point and keep it there. With a finger of the other hand, try touching circuit sections just as you did before, listening for changes in sound. This time, mark where both fingers were to create the sound so that a wire and body-contact can be connected to each.

The entire circuit can be searched in this way, with the stationary finger moving from point to point as the traveling finger checks the rest of the circuit.

The Screwdriver Approach

Here you'll do the same as just described except that you'll use bare metal jeweler's screwdrivers instead of your fingers (see Figure 9-12).

Do the same procedure—search first with one screwdriver, then with a screwdriver in each hand.

The advantage is twofold: You'll be able to isolate precise traces on the board that your finger-tip could only generally indicate (this happens when many traces run right next to each other), and the screwdriver tip can often get to circuit parts that you can't reach with your eager, though gigantic, fingertip.

FIGURE 9-12: Using screwdrivers to search for body-contacts

What Else Are You Looking For?

Now you'll pick up the bending wire you made and go deeper. This is where the real treasures (and dangers) lie. Let's take a look at what's possible as well as how to avoid trouble.

Floating Downstream

When electricity leaves the batteries, it does so at full force. But "downstream," as the electrical energy passes through different components, the voltage often gets "thinned down."

Take a good look at the battery compartment, as well as whatever circuit board traces it connects to, and try to avoid touching these areas during your initial bending tests, described in the next section. If you don't avoid them, you'll run the risk of allowing full voltage to flow through your bending wire into a circuit section downstream and operating on much less voltage. Result? You get fries with your Happy Meal whether you want 'em or not.

Exploring the Circuit, Step by Step

As described earlier, the basis of bending always follows the same course. Keep one end of the bending wire in place upon a single circuit point while the traveling end of the wire explores the rest of the circuit (avoiding, as said, the battery compartment and any wires or printed circuit traces coming from it). The stationary end moves to another point and the traveling end repeats its tour as you chart the responses. Here's the actual breakdown.

Preliminarily, be sure that the unit is running on fresh batteries and feel the circuit with your fingers as it operates to see whether any of the components heat up at all (voltage regulators and a few other items do warm up in normal operation). Have your charting materials ready. Have your bending wire at hand, your body-contact search completed and charted, and your safety glasses on.

1. With the instrument making a sound, place one end of the bending wire upon a printed circuit trace or metal component lead on the board, such as on the "leg" of a resistor or capacitor (see Figure 9-13).

FIGURE 9-13: The circuit-bending probe in action

2. Touch the other end of the wire, *very briefly*, to another circuit point. In fact, just briefly tap the traveling end to the chosen connection. Why briefly? Because right now you're looking for danger signs, as you will with each new connection. These signs include:

   - The circuit dies.

   - There's a loud pop from the speaker.

   - There's a loud hum from the speaker.

- You see a spark.

- Lights or displays dim.

All these signs indicate a possible circuit overload. If any of these things happen, abandon the connection and try to avoid making it again (you might mark it as a no go).

If the circuit dies, try reviving it by turning it off and back on. If this does not work, try removing a battery (to completely disconnect power from the circuit) and then replacing it. If this doesn't resuscitate the circuit, welcome to the bending graveyard and know you're in good company. Go get another toy and start again.

| Tip |

Even after such total cardiac arrest, a "dead" circuit can sometimes be revived! Using one hand, make your bending probe send voltage from the positive side of the battery compartment into ICs or transistors on the board. Just quick zaps here and there. If you're lucky, you may hear a little pop followed by a familiar demo tune! But why one-handed? So you can keep your fingers crossed with the other.

3. If there's no response, well, such is life at times. The traveling end's hike takes over and another circuit point is nudged, and so on through the night.

4. Mark all good bends on the board or in your notes for the next section, where you'll wire things up.

The Lucky Seven

With a little luck and your circuit-bending wire, you'll find the seven classic bends. You might not find them all in one instrument, but I'm sure all will show in time. They are looping, streaming, one-shot, the infamous "1 to X," potentiometer, LED, and photo cell bends. So, let me introduce each.

Looping Bends

As you search with your bending probe, you might find a connection between two circuit points that causes a sound to repeat over and over again. That's a looping bend.

To implement this bend, simply mark the two points and later solder a wire to each point. The other end of each wire will go to the two terminals of a toggle switch that you'll later mount on the instrument's case. If you need a refresher, see the toggle switch section in Chapter 6. Flipping the switch will initiate the looping.

Streaming Bends

Doing the same search as just described with the bending wire, you may make a connection that causes a stream of unusual sounds to occur as long as you keep the wire connected between two points. This is a source of aleatoric music (chance music, that is) because often these streaming bends string on-going unusual musical elements together.

You'll implement this streaming bend in exactly the same way as the looping bend: with a toggle switch.

One-Shot Bends

This is a little different. As outlined previously, the bending search begins with the traveling end of the bending wire only momentarily touching the next point on its tour. On some occasions this brief tap will launch the circuit into exotic behavior of one kind or another.

Again, great aleatoric music might result from this brief tap, the music continuing even after the traveling end of the wire is removed. Unlike the looping and streaming bends that require a continuous connection between two points to operate, one-shot bends need only that momentary contact to start-up.

To implement a one-shot bend, you'll once again solder a wire to each of the two points that created the one-shot response. But instead of a toggle switch, you'll need to solder the wires to the two terminals of a normally open (N.O.) pushbutton switch. Now you can just tap the pushbutton to get the same electrical reality (and sounds) that you discovered by momentarily touching the traveling end of the bending probe to the circuit.

The Stupendous "1 to X" Bend

Let's say that you have the stationary end of your bending probe on the lead of a capacitor and you get a nice effect when you touch the traveling end of the probe to a nearby resistor lead. Fine. You note the bend on the board with your pen.

Keeping the stationary end where it is, place the traveling end on another resistor lead and you get another cool sound. Great! And let's say that as the traveling end continues its search, you discover three more spots where cool sounds erupt. This is the "1 to X" bend, where "X" represents the variable number of responsive circuit points found in this way, with the stationary end of the probe sitting still on one spot. You'll find many "1 to X" bends as you continue to design instruments on your own. But how do you implement them? Super simple!

In the diagram (see Figure 9-14), point "1" represents where the stationary end of the bending probe was located during the "1 to X" discovery. Points "a"–"e" are the cool sound-producing areas that the traveling end of the probe revealed. You'll notice that a wire from point "1" is soldered to the middle lug of all five toggle switches—a switch for each of the five bends discovered. This puts all five switches in contact with common point "1."

Each of the switches is now connected to one of the five other points by means of a wire soldered to the switch's outside terminal. Just as you did with the bending probe, you have now made switch 1 connect point "1" to point "a." Switch 2 connects point "1" to point "b," and so on down the line.

Remember bus wire? That's bare wire without insulation to be removed. If you mount the switches in a "1 to X" bend in a row right next to each other (as in the diagram), you can use bus wire to connect all the common center terminals, the terminals that all go to the "1" point in the "1 to X" configuration (point "1" in the diagram).

How? Just start at one end and wrap the bus wire tightly around each terminal as the terminal is reached. Solder the wire to each terminal one after the other, on down the line. Now, using the regular wire-wrap wire, you can solder this line of switch terminals to point "1" in whatever "1 to X" bend you discover.

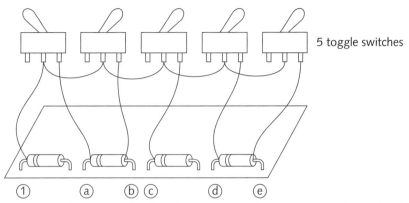

5 toggle switches

FIGURE 9-14: The "1 to X" bend. In this example, X=five, with each of the five bends running through its own toggle switch, the first of which is turned on.

Note

Because you were not able to audition these bends in combination with each other, having discovered them one at a time with the traveling end of your probe, you may have some surprises in store because with the switches, you can now turn more than one bend at one time. This capability creates electrical realities within the circuit that were not in effect during the exploration process. Although switches 1 and 2 might sound great together, switches 1 and 3 might be bad news for the circuit. If any trouble signs appear (see "Exploring the Circuit, Step by Step," earlier in this chapter), immediately turn off the offending switch. Your choice is to remember not to use that combination again or to find another "X" point in your "1 to X" scheme that does not present a problem when used in combination with the others.

Switches or Patch Bay for the "1 to X"?

The "1 to X" bend lends itself to patch bay wiring. A patch bay does the same thing as switches—it sends signals to your choice of destinations.

Figure 9-15 shows the previous "1 to X" bend, but configured as a patch bay.

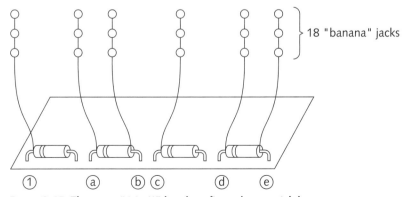

18 "banana" jacks

FIGURE 9-15: The same "1 to X" bend configured as a patch bay

You'll see that point "1" goes not to a switch but rather to a set of jacks (called banana jacks). As with the middle terminals of the switches, all "1" jacks are wired together in the "1" patch section of the bay. All thereby connect to point "1" on the circuit.

Points "a"–"e" are also brought forward from the circuit, but they also now go to sets of banana jacks, three (or more) jacks for each of the five points.

As you can see, plugging a banana cord (called a patch cord) between jack "1" and jack "a" does the same thing as switch 1 did before (see Figure 9-16).

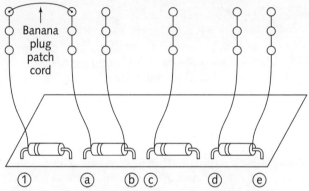

FIGURE 9-16: Patch bay making the same connection as the first toggle switch in Figure 9-14 ("1" to "a").

So, what's the advantage with a patch bay?

That's where the extra jacks come in (see Figure 9-17).

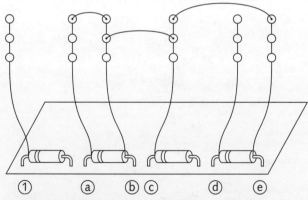

FIGURE 9-17: Patch bay making connections impossible with the toggle switches

Now there is no connection between "1" and "a"; instead, the connections exist between "a" and "b," "b" and "c," and "c" and "e." These, as you can see, are only a few of the new connections possible within the "1 to X" matrix that you could not implement were you using only switches.

So! For quick sound switching you can go with the "normalized" (referring to classic synth lingo) circuit of toggle switches. If you want more access and have the space to spare, the "non-normalized" patch bay might be for you.

Between the Wires

For this exercise you'll need to open another set of test leads. Cut and strip the wires just as you did with the first set. This time, *do* tin the stripped ends with solder: As in the exercise previously, twist the loose wires first into a firm spiral and then heat with a soldering iron until the solder melts easily into the twisted bare wire. Rather than solder the ends together to make one wire, solder an alligator clip to each tinned wire end (see Figure 9-18).

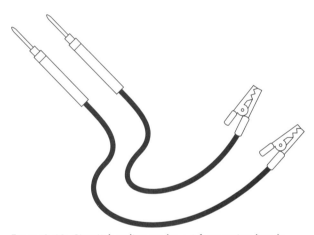

FIGURE 9-18: Circuit-bending probe set for running bends through various components

You'll use this set now for running bends through various electronic components by clipping the components between the two alligator clips. Now you once again have a bending probe, but it has whatever you want to test in the middle, causing the electricity being diverted to pass through the lucky whatever you've enlisted.

Pot = Variable Weirdness

Remember potentiometers: variable resistors controlled by means of turning their central shaft (Chapter 6)? Potentiometers are usually used for boring things such as volume controls. But in circuit-bending, all kinds of weirdness can happen as you throw pots into the mix.

First, any body-contact that required two points for operation is asking to be tried with a pot. Simply clip a pot between the alligator clips of the probe set you just made; then, place the probe points on the body contact points. Now the current will flow through the pot and the sound will change as you turn the pot's shaft.

But which pot? Try three. Try a 100K, a 1M (meg), and a 5M. As you learned earlier (again, Chapter 6), always connect one wire to the middle lug of a pot and the other wire to either outside lug (see Figure 9-19).

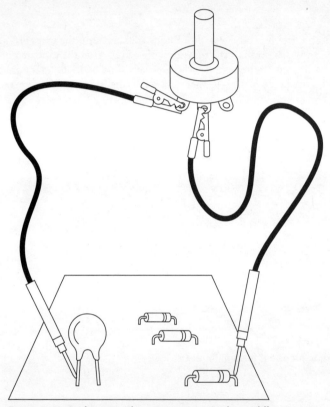

FIGURE 9-19: **Probe set with potentiometer in the middle**

Any looping or streaming bend you've discovered might as well be a good shot for a pot. Give it a try. If the circuit crashes while you turn the shaft, see the following "trim pot" section.

To discover brand-new potentiometer possibilities, just perform the usual circuit search with the pot in the middle of your probe set. Declare a stationary end and a traveling end and search away.

For the initial search, be sure that the pot's shaft is in the middle of its range. If you get a cool sound, stop and try to turn the pot's shaft to see what happens. Sounds as though you need a third hand for this, but I usually find a way to hold both probes in place with one hand while turning the pot's shaft with the other hand. Oh, did someone say art was easy? Whoops. You're right—I did. Okay.

You can make (or buy) two test leads with tiny alligator clips at the ends. If the clips are small enough, you might be able to clip them to some of the circuit sections, such as various component leads. With the clips in place on the circuit, you can then clip the pot between the remaining clips and handle the shaft with ease (see Figure 9-20).

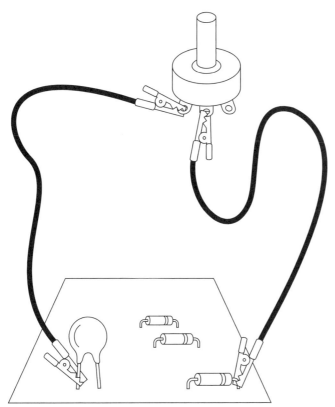

FIGURE **9-20: Leads with alligator clips at each end being used as the probe set**

Note   Soldering a short, stiff wire to circuit areas otherwise impossible to clip to (such as a printed circuit trace) will give you a custom terminal for your alligator clip (see Figure 9-21 and "Soldering to the Circuit Board" in Chapter 8). Search on!

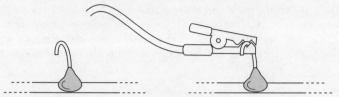

FIGURE 9-21: Soldering a wire to a hard-to-clip area will give you something to clip a test probe to.

Pots come with a caution. As you turn the shaft, you're varying the internal resistance between the pot's rating (its maximum resistance) and no resistance at all, which is what happens when the dial is turned all the way toward the outside lug you soldered (see Figure 9-22).

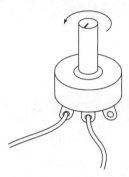

FIGURE 9-22: Turning a pot all the way toward the wired outside wired terminal reduces the pot's resistance to zero.

As you near the no-resistance-at-all-position, you must again be aware of the danger signs and cease turning the dial when any such trouble sign arises. Here they are again:

- The circuit dies.
- There's a loud pop from the speaker.
- There's a loud hum from the speaker.
- You see a spark.
- Lights or displays dim.

Let's say that as you turn a pot toward lower and lower resistance (again, turning toward the outside soldered lug), all is fine until you get to a certain point—and the circuit crashes. This means that the electronic flow through the pot at the lower resistance level was too much for the circuit: You let too many electrons through when the shaft neared the end of its rotation. What now?

This is what "trimmer" pots are for. The problem is that too little resistance is present in your new bend when the pot is turned too far. So, you need to find a way to allow the pot to be turned all the way without the decrease in resistance causing the circuit to crash. We need to find a way to be sure *there's still some resistance present* even with the pot turned all the way down.

Well, if you place a trimmer of the right value in line with the pot, then even with the pot turned all the way down the electricity still has to pass through the trimmer. And the trimmer supplies the resistance that the circuit needs to operate without crashing.

If you run into this problem, here's what to do. Let's say that you've found a good place for a 1M pot (remember to always try various pots in a pot-sensitive circuit). But as you turn it down, the cool effect, as mentioned previously, gets to a certain point and then crashes the circuit as you decrease the pot's resistance past a certain point. You need to "trim" your main potentiometer, and here's how:

1. Solder a trimmer of the same value (1M) to the full-sized 1M pot, center lug to center lug, and solder a wire to an outside lug of each pot (see Figure 9-23).

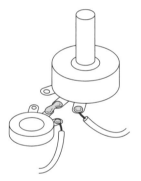

FIGURE **9-23:** Solder the pots' middle terminals together and solder a wire to an outside lug of each, as shown.

2. Turn the trimmer's dial to full resistance (away from an outside soldered lug) and turn the large pot's dial to no resistance (toward an outside soldered lug, see Figure 9-24).

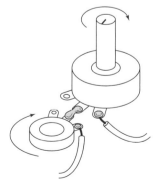

FIGURE **9-24:** Turn trimmer dial away from an outside wired lug; turn main pot shaft toward an outside wired lug.

3. Connect to the circuit and turn the trimmer's dial to just pre-crash and leave it there.

Now you've used the trimmer to establish a minimum resistance always present in the new bend. This is cool because you can now turn the larger pot all the way without risking a crash.

Note A trimmer of lesser resistance value than the main pot can be used in this scheme. If the trimmer's adjustment of the sound is too coarse, try trimmers of lesser values to fine-tune the pre-crash threshold.

Photo Cells

Variable resistors themselves, photo cells are those light-sensitive wafers discussed back in Chapter 6. You can probably use a photo cell wherever a potentiometer works. Rather than turn a dial to get the effect, with a photo cell you can just vary the light falling upon it. Try hand shadows!

The search process is just the same as with pots: Clip a cell in the middle of the alligator clip probes and see what you can find. Any time you get an interesting change in sound, do your best to cover and uncover the cell all the way so that you can hear the entire response range. Really great things can happen here!

Because pots and photo cells do the same thing but in different ways (one by knob, one by light), you might want to use both in the same bend, with the ability to switch between the two. This is a really simple wiring scheme and can be used to switch between all kinds of different components. Just substitute a pot and photo cell for the resistor and capacitor shown in Figure 9-25.

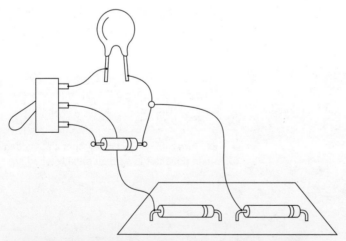

FIGURE 9-25: Simple circuit for using a three-lug toggle switch to choose between two components. This two-way circuit can be used for any two components, including the noteworthy pair of potentiometer and photo cell mentioned in the text.

Solar Cells: Power or Modulation?

The solar cells' role of converting light into voltage can be used to replace the batteries in an instrument. Each cell puts out about ½ volt in bright sunlight. If you solder enough together, doing so end to end as with batteries, you'll have a power supply—at least until the sun goes down (see Figure 9-26).

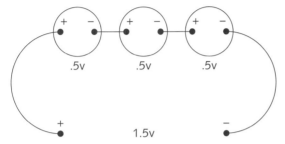

FIGURE 9-26: Solar cells wired end to end (in series) provide instrument power.

But you can also use solar cells as light sensors that behave differently under various lighting conditions. This can be interesting.

Explore as usual, but with the cell's leads clipped between the probe's alligator clips as you search the circuit. In some instances you'll be injecting a little voltage into the circuit (if the cell is in light). Other times, you'll be adding resistance to the circuit. This very unorthodox use of solar cells might ruffle NASA's feathers, but you're a circuit-bender and the Sea of Tranquility isn't what you're after.

Light-Emitting Diodes

LEDs are really cool, especially if you remember the days of trying to use tiny tungsten bulbs for pilot lights and such. LEDs require far less current, which allows them to light under conditions that would have tungsten lamps snoring in the tent. You'll use LEDs for pilot (power) lamps, envelope lamps (they vary with sound intensity), and logic lamps (indicating circuit activity).

The search process is the same as with the previous components, but with one difference. LEDs, as outlined earlier (see Chapter 6), are polarized components. If you were to "unplug" an operating LED's two leads and plug it back in with the leads reversed, it would not light.

To get around that here, you'll have to search the circuit twice each go-round: first with the LED clipped between the probes one way, and then again with the leads reversed.

Tip

If you have two identical LEDs, you can twist their leads together, anode to cathode, and clip them between your bending probe as usual. If your bend contains enough juice, regardless of polarity, to light the LED, one or the other will light. Note which LED lit, note its polarity, and now you *don't* have to search the circuit twice!

As usual, touch each new spot with the traveling end of your probe only very briefly. LEDs can burn out (even pop) if the current is too high. If an LED lights brightly for a split second and then glows much more dimly (and sometimes off color), you may be about to burn it out.

Sometimes an LED will act similarly by being under- rather than overpowered. But an underpowered LED does not change color; it only shines dimly. An underpowered LED does not get hot the way an overpowered one surely does prior to burning out.

Pilot Lamps

Many toy circuits don't have a pilot lamp and don't turn themselves off. Adding a pilot lamp is a nice addition to this type of instrument. And pilot lights are cool, anyway.

As I said, search the powered-up circuit with the LED clipped into your probe. Don't worry about sound this time. Just be sure that the unit is turned on.

Try different LEDs in your searches. Regular, high-brightness, small, large, and all colors. When your LED lights and seems to be getting the right voltage (glows nicely, does not warm up very much), turn the circuit off.

If the LED goes out, you've found your pilot light. If, on the other hand, the LED remains lighted, you've discovered a live circuit area even with the power off. Weird, eh? Keep searching.

After you've found a spot that lights the LED, and the LED goes off when the power is switched off, you're almost there. The last thing to do is to test the circuit with the LED lighted. If all goes well and the circuit operates as it did before the LED was added—you're done!

Logic Lamps

I'd say these are just eye candy, like the manifold vacuum gauge on my 1970 Pontiac Grand Prix. The car would have worked fine without it looming there, right next to the Craig 8-track player on the windswept console between the deep bucket seats. But it did tell me conditions within the manifold as I hit highway speeds, the red zone translating directly into "you're now getting 3 miles per gallon."

Anyway, get a sound going so that you have some circuit activity happening. This time, look for the LED's light to fluctuate as you go from spot to spot with the traveling probe. With luck you'll find a few spots where this will happen, perhaps flashing the LED with no seeming connection to the sounds. That's a logic lamp. Once again, be sure that the LED goes out when the main power switch is turned off and that the circuit is not affected adversely by the LED's introduction.

As with my old manifold gauge, mostly eye candy. But logic lamps *do*, as does a manifold gauge, tell you things about your instrument that you didn't know before. And in that way they can be, I hesitate to say, illuminating.

Envelope Lamps

The envelope of a sound is often described by its volume event. A bell's volume event, or envelope, is immediately loud at first and then quickly drops in volume, finally fading away. An LED whose brightness follows the volume envelope of a sound is an envelope lamp. Bright when the sound is loud; dimmer when the sound is low; always changing. Look at it as a VU meter with a personality disorder.

Another kind of envelope lamp you'll find now and then might better be called a peak lamp. As with the overload indicator on recording consoles, it flashes only in the presence of higher current.

Look for these in the same way as before, with the circuit making a sound. But start at the speaker terminals. Use a *high brightness* red, blue, or white LED and remember to reverse the LED's leads to allow for the LED's polarized design.

"I Didn't Know It Could Sound So Good!"

That's exactly what you'll say when you hook up your alien instrument to an amp for the first time. People, you know, are quick to harass circuit-benders for working with audio from toys, thinking it's gotta be strictly lo-fi. Nay.

To run your interociter (no? watch *This Island Earth*) into an amp, you need a line output. And to get a line output, all you need to do is solder a wire to each speaker terminal.

I'd like to tell you that there's a standard color code for positive and negative speaker leads. Red may be positive. Black may be positive. But then, yellow might be positive, blue might be positive, even green or violet might be positive. The result? Anything can be positive—except you.

Look closely at the speaker connected to the circuit. You should be able to see small "+" and "−" signs on the plate that holds the speaker's terminals. The wire that you solder to the "+" terminal goes to the positive terminal on the jack that you'll install later. You'll remember that the positive terminal on a jack goes, by means of a springy contact arm, to the tip of the plug you'll insert.

The "−" wire will go to the remaining terminal on the jack, the "common" or "ground" terminal connecting the outside ring of the jack. This outside ring of the jack is meant to contact the shaft or "barrel" of the plug when inserted. Need a jack refresher? Review in Chapter 6.

At times the signal from the speaker will be too hot (loud) to serve as a good line output. The answer is a trimmer pot, as shown in Figure 9-27.

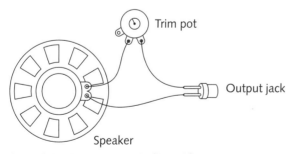

FIGURE **9-27: Put a trimmer in-line with an output wire to tame your line outputs.**

Use a 500Ω (but experiment with various resistance values) trimmer in line with the positive side of the jack. All that's left is to adjust the trimmer to a lower output level. Give it a try.

The Panic Button

You probably have something with a panic button on it—that reset button hidden on your camera, for example. Those tiny, shame-inducing little holes you have to stick "a sharp object" into to bring flipped-out digital freeze-ups back to reality again.

What the heck's actually going on here within these high-tech, super-refined, cutting-edge electronic oracles when they get their belly buttons pressed? Well, even with their true-theory electronics, they've crashed just as your bent Speak & Spell has, and they need the exact same remedy. When you push their reset buttons, you disconnect their battery supplies for a moment, and often nothing more is needed. It's the same as taking the batteries out and putting them back in, just easier.

But why not just turn the instrument off and back on again? Nice thought. Just might work, too. But in some cases a nonoperative state in the circuit remains in effect, even with the power switch turned off and back on. The only solution here is to remove the batteries and put them back in, or do the equivalent with a reset switch: interrupt the battery power supply.

Not all bent instruments need a reset switch. If the instrument you're working on tends to crash, and the power switch is a hassle to use or it doesn't do a full reset, the pushbutton reset is the answer.

For this you'll need a pushbutton switch. But this time you'll need the less used "normally closed" (N.C.) pushbutton. Remember? A normally closed pushbutton breaks the circuit when pressed (see Chapter 6).

If the battery compartment is connected to the circuit with wires, you're in luck. Simply cut either wire and solder your normally closed pushbutton into the gap (see Figure 9-28).

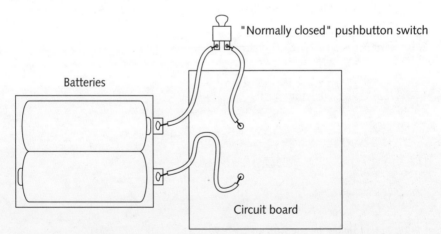

"Normally closed" pushbutton switch

Batteries

Circuit board

FIGURE 9-28: A normally closed pushbutton switch on a battery compartment wire

If, however, the battery compartment is connected to the circuit directly via traces on the printed circuit board, you'll have to do a little surgery. No big deal.

Look closely and locate a trace that leaves the battery compartment and enters the circuit. Find a place along the trace where you can cut it prior to the trace's connecting to other components

Your job is to cut the trace in a way that leaves enough of the trace on either side of the cut to be soldered to (see Figure 9-29). You might want to review the info about soldering to PCB traces again; for that, see Chapter 8.

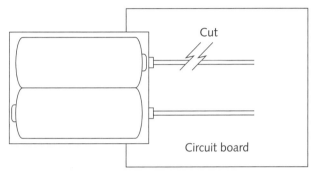

FIGURE 9-29: Cut power supply trace.

Cut the trace slowly and carefully by scraping with a sharp X-ACTO–-style blade, or use a tiny burr bit ($\frac{1}{8}$" works well) in your Dremel drill. When you're through the trace, you'll see the circuit board material beneath. Solder a wire to the trace on both sides of the cut. These two wires will go, of course, to the two terminals of your normally closed pushbutton switch (see Figure 9-30).

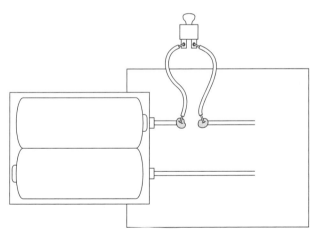

FIGURE 9-30: Solder a normally closed pushbutton switch across the cut trace.

When the switch is at rest, it allows electricity to flow around the cut. It flows through the normally closed switch instead. But when the switch is pressed, it breaks the flow from the battery compartment going through it. That's about as deep a reset as you can get.

Building the Bent Instrument

Weird how rumors travel. Especially within circuit-bending. No, flexing a circuit board is not circuit-bending. And unless you want to send human tool use back into the Iron Age, hot soldering irons, no matter how well you personally metabolize polystyrene smoke, are not drills. Let's get things right.

Case Preparation

That plastic instrument housing is just waiting for all the new controls you need to add to implement all the bends you've discovered. It's yearning for 'em. But you may have both problems and assets hiding right under your nose.

Clearances

So, you've painted your alien instrument, installed all the switches, all the pots, and all the dials and LEDs. And it looks great! You carefully line up the halves of the case, bring them together, and—what? Why don't they go together? They don't go together because the backs of the switches, the pots, and the dials and LEDs all hit the circuit board. Bummer.

To avoid this all-too-common experience, *be sure* to look before you leap! Try to locate all your added components at points on the case that will not interfere with potential obstructions, such as the circuit board example mentioned, when the instrument is reassembled.

If you do want to mount components opposite the circuit board or anything else within the case, just be certain that there's still room. To do this, measure the depth of the components that will be present on the backside of the panel they're mounted on.

If the backside of the mounted switch measures, say, ½", just tape a ½" stack of pennies to the spot on the inside of the case where the switch would mount (see Figure 10-1).

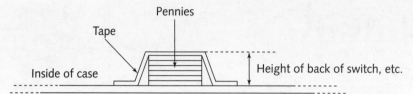

FIGURE **10-1: Stack pennies to height of back of component and tape to inside of case.**

If the case closes with the pennies in place, you have pretty good assurance that the switch will fit. Remove the pennies. Install the switch.

It is a good idea to add a couple pennies to the stack to make sure that you'll still have clearance after wires are soldered to the switch.

Rib Surgery

If the circuit board is in the way, you're pretty much stuck with that and you'll have to plan around it. But elsewhere, you might have options.

Many plastic cases are made more rigid by means of "ribs"—internal plastic walls that criss-cross between the case's outside dimensions. Removing these ribs usually has little effect upon the integrity of the housing. If such ribs block the placing of components opposite them on the other half of the case, just clip them away with a high-quality wire clipper (the soft plastic won't harm the clipper). A small pliers can be used in conjunction with the clippers by twisting sections of the clipped rib away from the case.

Finalizing Component Locations

Two things come into focus here: clearances and playability. Observe the positions your hands fall naturally into when touching the instrument. Look for comfortable hand and finger positions and, keeping internal clearances in mind, plan your control array as close as possible to these hand and finger spots. However, do the opposite for your reset (power supply interrupter) switch. Find a place to mount it as far away from your finger positions as possible). This is a switch you don't want to hit by accident!

Using a pencil (because it's easy to erase with a cloth or fingertip), mark *very lightly* where the components will go. If you want components to be spaced equidistantly, measure. Using a ruler, make layout lines drawn in very light pencil to keep components in a row (see Figure 10-2).

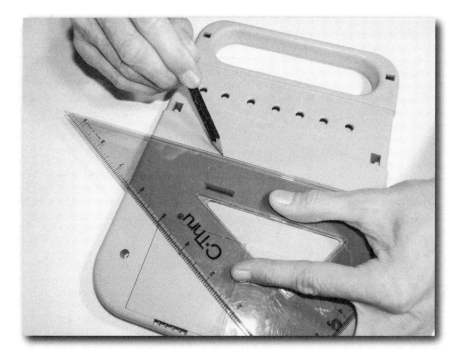

FIGURE 10-2: Use a ruler to lay out straight lines.

Holding the edge of a plastic ruler in a curve will help you draw smooth arcs (see Figure 10-3). Don't forget the mechanical compass for drawing arcs, either.

FIGURE 10-3: Use a flexible plastic ruler for drawing curves.

Drilling

Good drill usage starts with realizing that you + drill = human drill press. But whereas a drill press looks to a rigid metal frame to keep things in line, you're but flesh and blood.

Think of the rigidity of the drill press as you use the Dremel hand tool and try to replicate the press's framework by keeping your torso and arm muscles somewhat firm. Missing this, you're likely to allow the spinning drill bit to run across the plastic from time to time, leaving an undulating scratch along its path.

In the event of drilling through flat panels (such as project box housings), back the panel with a wood block of some kind to avoid bending the panel or having the bit hit your workbench upon penetration.

Have your workpiece well supported, and if you need for some reason to drill a hole with an unmovable item behind it (that is, circuit section), be sure to control the bit well so that upon penetration it does not contact any nearby components.

Painted Holes Are Different

A well-painted instrument might have five or more coats of paint after base coats, top coats, and final glosses are complete. These layers may build up to $1/32$" or more at the inside edges of holes. This will reduce the diameter of holes by $1/16$" or thereabouts. The answer, of course, is to make holes a little larger than usual if the instrument is to be painted. You'll have to get used to the exact adjustment needed relative to the paints you use.

This entire issue becomes important only when the hole edges will be seen, as in mounting an unhoused LED or photo cell for which fit and finish should be exact. In less critical situations, you'll be able to remove the paint from the inner edges of the hole with a hand bore (as long as the edges of the hole will be hidden by the mounted component's collar, as is the usual situation with a light or other fixture designed for panel mounting). I discuss hand bores in a moment.

Bits

The most common drill bit is the spiral bit. A newer version is the self-tapping spiral bit. It looks like its simpler namesake but with the addition of a smaller protrusion at the tip that is meant to drill a pilot-like hole before the main bit hits the work material. Either will work fine for most of your drilling requirements (see Figure 10-4).

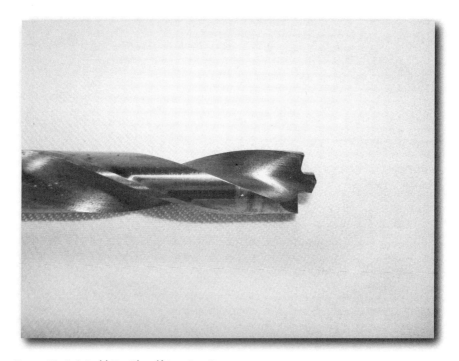

FIGURE 10-4: Spiral bit with self-tapping tip

Mentioned previously, burr bits are bits with rasp-like balls at the tip (see Chapter 5). These are used for carving as well as opening pilot holes to their final diameter. But look to a hand bore for hole enlarging.

Spade bits, made for wood boring, can also be used on thicker plastics and Lucite. These larger bits are to be used with a hand drill or drill press, not a Dremel drill. But if your projects need holes wider than your spiral bits and hand bores provide, spade bits can be quite handy (see Figure 10-5).

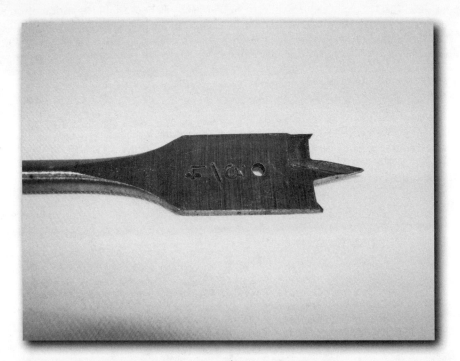

FIGURE 10-5: Spade bits can be used, with caution, on thicker plastic.

If you're using a spade bit in a power hand drill, be sure to remember to keep your body frame rigid and under control, *especially as the bit finally breaks through the material*. Also be sure to have the workpiece backed by a wooden block if possible. This will keep the bit on track and should minimize damage to the workpiece (spade bits are not known for their gentleness— practice on scrap material first!).

Note   Spade bits are *not* for the usual thin plastics of circuit-bent toys. Reserve them for materials thicker than 3⁄16", such as custom Lucite enclosures and control panels.

Are you working with metal cases? You may want to use a *center punch* (see Figure 10-6) to indent the positions of the pilot holes before drilling. If the metal is soft, like aluminum, be sure to back the metal with a block of some kind so that the punch doesn't dent the metal beyond the small indent you're after.

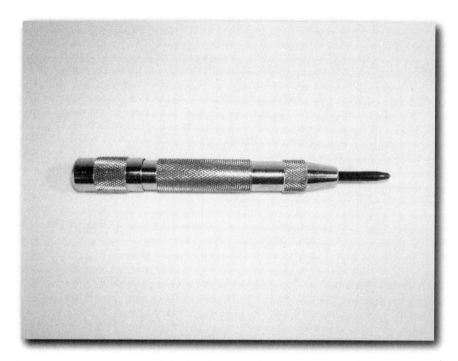

FIGURE 10-6: Use a center punch for locating holes to be drilled in metal.

Bit Entrance and Exit Sides: Big Difference

When a drill bit (especially a spiral bit) enters the work material (most often plastic in our examples), it usually leaves a rough hole edge behind. But when it exits the opposite side of the material it usually leaves behind a super-clean hole.

This is important to know if the hole will be seen and not blocked by a component collar (such as some pilot lamp housings) or component hardware (such as the flat hex nuts of toggle switches).

Let's say that you want to mount a LED right on the case without placing it in a housing of any type. If you mark the outside of the case and drill a hole at that spot from the outside, you'll have the "dirty" side of the hole staring you in the face. Solution? Drill the hole from the inside out. Now the LED will enter the rough side of the hole on the inside of the case and exit the smooth side of the hole on the outside of the case.

Practice Holes

In the preceding example of the LED, you'll want it to fit snugly into the hole. To test this fit (and others), keep a piece of drillable scrap material at hand. Choose a bit that looks to be the same diameter as the LED. Drill through the scrap plastic to test the LED's fit.

Pilot Holes

A pilot hole, of course, is the small hole you drill before the hole is brought up to component size. Drilling pilot holes makes drilling the actual hole size you need much easier, as well as allows you to reposition the hole later. I use a ⅛" spiral bit for most pilot holes.

Hole-Drilling Fixes

Spinning drill bits are wanderers. Should you not have them positioned truly perpendicularly to the workpiece, upon contact their spinning edge wants to roll along the workpiece like a wheel. Keeping them centered and on the mark can be problematic.

Then there's the unwanted hole, drilled for a component no longer needed. It's exactly where it was supposed to be. But it's for the switch that—whoops—turned your Speak & Spell into an electric deep-fat fryer. Let's look at fixing these two common problems.

Correcting Alignment

Provided that you've been using pilot holes, you can usually move a hole back on center. Not as magical as it sounds, and a real problem solver because everyone faces this from time to time.

Using the side of the spinning drill bit, you widen the off-location hole toward the correct position until its width is equal on both sides of the layout line (see Figure 10-7).

This rough but recentered hole is then brought up to the correct shape and size, along with all the rest, with a hand bore.

Filling the Unneeded Hole

Here you have a predicament much like that involving the paint drip, except that you can't sand a hole away. Or can you?

If you visit a good hobby or model shop, you'll find compounds designed to fill holes in plastics. These, along with some sanding, will easily plug the hole prior to painting.

But what if you've painted before you realize that the hole is unneeded? Now you're exactly in paint-drip territory, and all the same solutions apply. See Chapter 12 for some sneaky fixes.

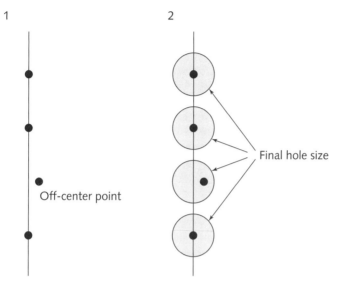

FIGURE 10-7: An off-center pilot hole can be enlarged back onto center.

Hand Bore Means Precision

The reasons I like to use a tapered hand bore instead of larger bits to bring pilot holes up to size are many. Your Dremel, even with its larger burr bits attached, can't make large enough holes for many of the components. A hand bore can.

Using a larger hand drill and big bits is unwieldy and can result in instrument damage. The hand bore is light and easy to control.

Bits are manufactured in stepped sizes. This means that if you use bits for the final holes, some components will not fit snugly in the holes. This small amount of play can hasten the loosening of components. A hand bore assures exactness in hole diameter as long as you're careful and work in small steps.

To use a hand bore, just insert its tip in the pilot hole and rotate the tool. As you widen the hole, keep trying the component for size so that you don't over-bore the hole and end up with one that's too wide (see Figure 10-8).

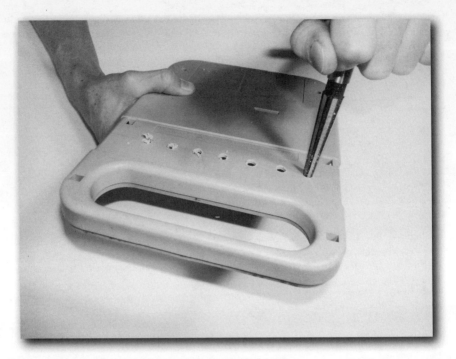

FIGURE 10-8: Using a hand bore to increase pilot hole size

De-Burring

The companion tool to bits and bores is the de-burring tool. Similar to the hand bore but with a faster taper and wider diameter, the de-burrer removes plastic and metal "burrs," the ragged material often left behind at the edge of drilled and bored holes.

As with the hand bore, you place the de-burrer against the hole and rotate it. This takes only a moment and leaves the hole's edges clean and flush to the surface (see Figure 10-9).

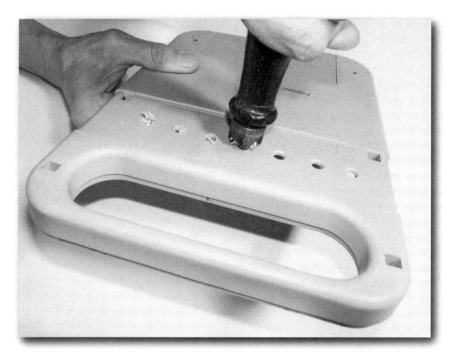

Figure 10-9: Use a de-burrer to smooth hole edges.

Mounting New Components

If you've decided not to paint your instrument, you'll be better able to survive tool slippage. If you've painted your instrument, incorrect tool use can lead to disaster: The tool slips and the paint gets scratched. The key to most of your component mounting is good tool fit. From my personal school of trial and error, here are some tips.

Switch and Potentiometer Hardware

Both of these components use the same mounting hardware: hex nuts. These screw onto the component's threaded collars. Resist the pliers temptation and reach instead for the crescent wrench set. Switches' hex nuts can also be tightened with your socket drivers, making the work even faster.

Tip Finding thin, good-looking metal washers to slip over threaded collars of switches, pots, and other nut-mounted components is a good idea. This way your wrenches or hex drivers will rest against the washer rather than the instrument while tightening, keeping the instrument's finish in good shape.

Potentiometers may come with long shafts. These are meant to be cut down with a hacksaw to the length required (just enough to go up into the knob you want to use). Clamp the shaft in a vise for making the cut (see Figure 10-10).

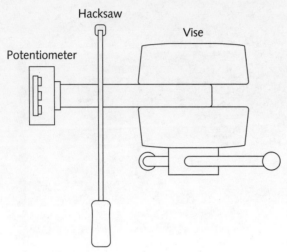

FIGURE 10-10: Use a hacksaw to shorten potentiometer shafts if needed.

Output Jacks

Guitar jacks and their smaller knock-offs are mounted with hex nuts, just the same as switches and pots. Grab the crescent wrenches or socket drivers.

RCA (phono) jacks are also mounted with hex nuts, the only differences being that the nuts screw on from the backside of the panel on which the jack is mounted. Many times you'll be able to use only crescent wrenches here because space limitations will knock socket drivers out of the running.

Body-Contacts

The brass ball style body-contacts you'll be using in the project section are threaded to accept a bolt. The bolt heads will be either slotted or Phillips depending which you bought to fit the body-contacts. Personally, I prefer Phillips head body-contact bolts because Phillips drivers are less likely to slip during use.

LEDs and Pilot Lenses

If you have a drill bit that's the same diameter as the LED you're mounting, you can just drill a hole and insert the LED from the inside side of the case (remember to drill the hole from the inside of the case because the exit side of the hole is cleaner).

If you're mounting the LED in a pilot lamp housing of some kind you'll probably again be dealing with hex nuts on either the inside or outside of the case. A crescent wrench or socket driver will do the job.

Increasing the Amperage and Life of Batteries

To put it simply, the difference between a AAA battery and a D battery is that the D will give you the same 1½ volts as the AAA, but for much longer. If the instrument you're using runs out of power quickly on its power supply of four AAA batteries, substituting four D batteries will extend play time considerably and usually is not too difficult a process.

Approach #1: The Convenient Auxiliary Power Input

If your instrument has an input for an AC adapter, things couldn't be easier. Obtain a battery compartment that houses the same number of batteries as the battery compartment built into your instrument, but one that houses larger batteries. D's can be substituted for AAA, AA, and C batteries (all supply 1.5 volts). If you're trying to substitute for a 9-volt battery, you'll need a battery compartment that houses six 1.5-volt batteries, preferably D's if you want the longest play time.

The battery compartment will have two wires coming from it, red for positive and black for negative. These you'll solder to a power plug that fits the AC adapter input already on your instrument. Here's what you do:

1. Take the instrument to the store you'll be buying the power plug from and be sure you choose one that fits exactly.

2. Look at the instrument to see whether the input is marked with a polarity indicator telling you whether the tip (center) of the power plug is supposed to be positive or negative. If this is not indicated on the instrument, it may be marked on the body of the AC adapter itself, should you have it handy (the tip is usually positive, though you can't count on this).

3. Keeping in mind the polarity, solder the wires from the battery compartment to the power plug (see Figure 10-11).

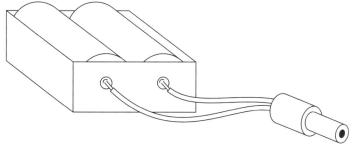

FIGURE 10-11: Battery compartment with power slug soldered on

Thread the wires through the plug's cover before soldering.

If the packaging of the power plug does not indicate the polarity of the soldering lugs on the plug, look closely and see which lug goes to the center of the plug and which goes to the outside of the cylinder, or "barrel," of the plug. The longer contact is usually the outside, or barrel contact; the shorter contact is usually the tip contact.

4. Now plug the battery-filled compartment into the instrument and see whether it works. If it does not, *immediately unplug the battery pack* and study the wiring to be sure that you wired the polarity correctly. If you were in doubt of the wiring to begin with (no polarity indication could be obtained from either instrument or AC adapter), try reversing the wiring going to the power plug and retest.

Approach #2: The Workaround

Often you'll be working on instruments that don't have a power adapter input. All is not lost. You'll have to wire the battery compartment to the actual circuit using your own, newly installed power input jack.

1. Obtain and install a panel-mount power input jack that will fit on your instrument (see Figure 10-12).

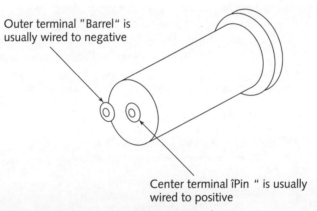

Outer terminal "Barrel" is usually wired to negative

Center terminal îPin " is usually wired to positive

FIGURE 10-12: Power jack soldering terminals

2. Study the battery compartment built into the instrument and find where you can solder to the existing power supply circuitry.

a. If the built-in compartment has two wires coming from it, solder two more wires to where these original wires connect to the compartment or the circuit (see Figure 10-13).

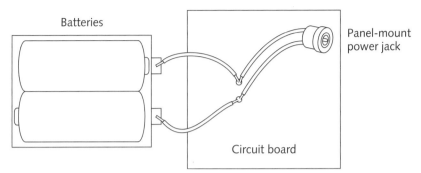

FIGURE 10-13: Solder extension wires to the wires already coming from the battery compartment.

    b. If the built-in does not have wires but instead connects directly to your circuit board, you'll need to solder two new wires to circuit points—one to the positive side and one to the negative side—of the existing battery compartment circuit traces (see Figure 10-14).

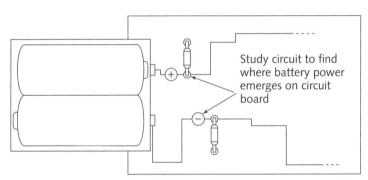

FIGURE 10-14: Extension wires will be soldered to components (or printed circuit traces) connected directly to the batteries (power supply).

3. Solder two power extension wires to the new power jack: positive wire to the lug that goes to the center, or "tip" connector, of the jack, and negative to the other lug (see Figure 10-15).

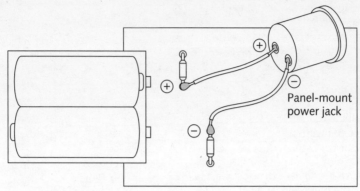

FIGURE 10-15: Noting polarity, solder extension wires to the power jack.

See packaging, if possible, to determine polarity. As mentioned, the terminal closer to center is usually positive.

Remember that in any jack, power or audio, for which a soldering lug obviously connects to the outside of the jack's structure (the part that will come into contact with the panel it's to be mounted on), this is the negative ("–" or "common" or "ground") terminal. The other terminal will always connect to the tip ("+" or "positive" or "hot") contact of the jack.

4. Last, wire your auxiliary battery compartment the same as in Approach #1, keeping polarity consistent.

With this approach, be sure to use either the built-in battery compartment *or* the larger auxiliary that you build, but *never* use both at the same time. If you're using the auxiliary compartment, *be sure* to remove the instrument's internal batteries first.

Note There are power jacks that interrupt the internal power supply's circuitry when the power plug is inserted. If you carefully follow the wiring instructions that accompany these jacks, you'll be able to leave your internal batteries inside the instrument when plugging in the external battery compartment. Just be sure that the internal battery supply is truly interrupted when the external compartment is plugged in. To test this, plug the *empty* new compartment into the instrument working on its internal batteries. You've wired the power-interrupt jack correctly if the instrument goes dead when the external battery compartment is plugged in. If it does go dead but does not operate when the external battery compartment is then filled with batteries, it could be that the polarity of the plug on the new compartment needs to be reversed.

Batteries vs. AC/DC Adapters?

In the world of power supplies, AC adapters are problem makers. These "wall warts" are not only layabouts but also ne'er-do-wells and bother boys all at the same time. Sorry to report that.

When you're bending a circuit, you want no connection to the "mains" whatsoever. The only thing keeping you from the 110 volts jumping around inside that black lump is a flimsy bit of insulation. That's not enough of a safety margin for the hands-on procedures of bending, a process that in itself might put pressures upon the AC adapter, causing it to overheat and malfunction. But there are more reasons to abandon AC adapters than this scary prospect.

Batteries are known as the cleanest power supply commonly available. Clean? This means that you're getting pure DC without the "noise" inherent in AC to DC adapters. These adapters must filter all kinds of garbage out of the incoming 110 volts of alternating current (AC) to finally supply you with the low direct current (DC) you need. Few perform this job well. Power supply noise dirties audio signals and, additionally, is hard on electronics.

Worse than dirty, many AC adapters are poorly regulated: Although their internal voltage regulator is supposed to output 6 volts, it actually puts out, say, 7.5 volts. This over-voltage can take a toll on the electronics designed to see less voltage. We're talking early instrument burnout.

I'm not done yet. Another regulation issue has to do with spike handling. Line voltage, the 220 volts entering your house, is subject to great voltage fluctuations, or spikes. These are transferred to the 110-house current "split" from the 220, and the spikes end up in your "wall warts," your lowly AC adapters. If poorly spike protected, the adapter will then slam a real jolt into your instrument. Now we're talking immediate instrument burnout.

So batteries it is. Get a few sets of NiMh and get used to recharging. Always buy the batteries with the greatest amperage available (they'll last longer).

In shopping for bendable circuits, you'll collect a lot of batteries. Many will be dead, but some will be good. Keep the good ones and dispose of all of them properly via a recycling station so equipped.

If you're stuck buying batteries, alkaline will beat the cheaper varieties. I've found major brands to compete pretty well. But I've also found new batteries leaking in their packs. Look closely.

Duracell is in the habit of using a separate metal plate for the positive tip of its batteries. This circular plate is seemingly held onto the rest of the battery only by the plastic name-printed sleeve that wraps the cylinder of the battery. Shoddy! Benders are constantly inserting and removing batteries. This wears away the plastic wrapper, the positive plate falls off, and your new batteries are worthless. Don't you hate to see this kind of poor work in the electronics industry? I sure do.

Bending Beyond the Basics

No one in the gallery had ever before heard anything like the strange sounds coming from behind the partition in the corner. Streams of notes were lilting up and down unusual scales, angelic yet fractured—like a sweet dream about to go bad.

In front of the partition was a pedestal. On top of the pedestal were two hand-sized metal foil pads. Curious patrons would finally get around to touching these pads, and soon everyone in the gallery would be gathered around.

When the pads were touched the sound leapt upward from the previous tranquility of mellow scales to wilder-than-wild lead work, all playable via simple hand pressure. Speculation was hot as to just what was hidden behind the scrim. Arp? Moog? Some kind of hands-on theremin?

Good guesses. But it was only my original, short-circuited transistor amplifier, the miniature version controlled by body-contacts and bought at a cost of $2. I'd run its output into an effects unit and that into an amp. It was the extension of two of the instrument's contacts out into larger pads atop the pedestal that transformed the device into the very mysterious instrument that visitors were trying to understand. A little aluminum foil can go a long way.

Component Relocation

Relocating the body-contacts as mentioned was an aesthetic decision. But down-and-dirty, nitty-gritty *need* is more often behind relocating components.

The classic example is the need to mount switches on the instrument case but not being able to because the backs of the switches will hit something soldered to the circuit board—a large capacitor, for example. Solution? Relocate.

1. Snip the capacitor (or whatever) from the board, about ⅛" from the solder junction itself, leaving on the board just enough remaining of the component lead still able to be soldered to (see Figure 11-1).

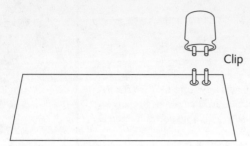

FIGURE 11-1: Clip the component from the circuit.

Observe when you snip the leads which lead went where just in case you're dealing with a polarized component.

2. Find a spot within the instrument case, as close to the original location as possible, to fasten the relocated component where it will now be out of the way (a little dab of hot-melt glue is good for this; see Figure 11-2).

FIGURE 11-2: Glue the component in an out-of-they-way spot.

3. Solder the wires between the component's leads and the spots on the circuit board where the component was originally located. This should be easy because you left little wire stubs behind when you first clipped the leads (see Figure 11-3).

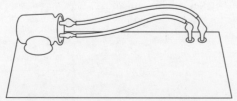

FIGURE 11-3: Reconnect the component by soldering wires between it and its original leads.

Note If you're very comfortable with soldering and don't need to leave wire stubs behind to solder to, either clip the leads flush with the board or heat the solder joints and remove the component, leads and all, from the board. Doing so will reduce the chance of the heated stub from the clipped lead popping out of the reverse side of the board and possibly causing trouble.

Who Needs an Extension Box?

You do, if you find so many bends that you don't have enough room on the original case to mount all the new controls. Using anything from a "project box" (empty plastic or metal case for building electronic projects into) to an old toaster, you'll find that expansion is close at hand.

Mount your needed switches, pots, LEDs, and whatever else on the toaster; then, connect all the wires to the circuit as usual. The real issue is *all the wires* that run between the circuit's original housing and the added box.

The easiest way to do this is with multiconductor wire and rubber grommets. Here's how:

1. Determine how many wires will have to run between the extension box and the original instrument.

2. Choose a multiconductor cable that supplies the many inner wires and cut it to the length you want between the instrument and the box, including the length of the wire "runs" inside both units. Leave a few more inches than you think you'll need.

3. Remove the outer plastic of the multiconductor cable to a distance from the ends as needed for the longest wire run expected within each unit, for both the original case and the extension box.

   For example, if you want 12" of cable between the two units and you feel the longest wire run inside each unit will be 10", start with a cable 3' long. You'll want a little of the outer insulation still intact inside each case, so leave 1" beyond the 12" you want between units. This means that you'd be stripping 11" of the outer insulation off each end of the wire (start with a long scoring of the wire with an X-ACTO blade along the 11" length to be removed, tearing along the score to remove and finally clip this sheath away at the 11" mark).

4. Strip ¼" off the end of each of the inner wires you intend to use to make your connections. Tin the wire ends (see Chapter 8, Figure 8-21). The cable is now ready (see Figure 11-4).

5. Obtain rubber grommets that will allow the unstripped cable to snugly pass through, and mount these within the holes drilled in the case and extension box (slowly enlarge pilot holes with your hand bore to achieve the correct hole size for the grommet's mounting diameter).

   Pass the cable through the grommet in the original instrument's case so that the still-insulated main cable just passes through. A nylon wire tie pulled tightly around the wire, just inside the case, will keep the cable from slipping out (see Figure 11-5).

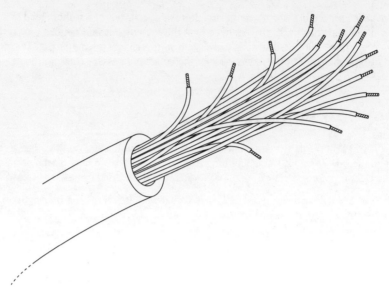

FIGURE **11-4: Strip away outer insulation; strip and "tin" the ends of the wires inside.**

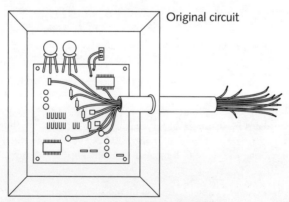

Original circuit

FIGURE **11-5: Insert cable through grommet and solder wires to the bending connections.**

6. Hot-melt glue will also secure this cable if applied over the nylon wire tie and up against the inside of the case.

7. Solder the inner wires to the bending connections on the circuit board that you want to access on the extension box.

8. Similarly, pass the other end of the cable into the extension box. Secure as before, and watching closely the wire color codes, solder the wires to the extension box components just as you would if the switches and so on were mounted on the original case itself (see Figure 11-6).

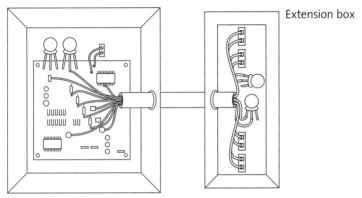

Extension box

FIGURE 11-6: Insert the cable's other end into the extension box and solder wires to extra components.

Rehousing a Circuit into Another Case

I really like the aesthetics of the transformed instrument, with the original housing still there but bristling with new controls. In other words, I shy away from hiding the source of a bent instrument by putting it inside another case; I prefer the more tangible metamorphosis of additive adaptation, with controls growing from the body of the original instrument itself. But that's just me, and I'm not beyond question.

Still, there are times when there's not enough room to mount all the new controls, as discussed, or even when the original housing is just too unsettling to retain. (See "The Harmonic Window," in this chapter, for an example of both issues; this housing contains a sampler from inside a way-too-cute teddy bear.) A new housing for your bent toy might be just the thing.

Simple circuits, whose components can be replaced with your own, can be mounted in another case without too much trouble. Some of this transference of controls can be complicated, but if there are only a few pushbuttons (such as on many toys that play samples), these can be replaced with your own by just soldering wires to the contacts of the original switches on the circuit board and soldering the other ends of the wires to your own pushbuttons mounted on your own case, a case that can also accommodate all the new bending controls you need to add.

The circuit board is removed from the original case along with the speaker and battery compartment. If the battery compartment was integrated into the original case, you'll have to wire a separate, stand-alone battery holder to the circuit where the old one was connected. If the original power switch can't be extended to a new switch on the new housing, just leave the original power switch turned on and solder a toggle switch in the middle of one of the new battery holder's wires. This becomes your new power switch. Cool.

The "Project Box" Approach

Project boxes come in endless designs. These are made for experimenters needing a case for their DIY E-Bomb circuits, Royal Rife power supplies, and dummy pirate radio transmitters (don't give up the real one!). The best cases open easily for battery access, are nonflexible (sturdy), and have a little more room than you think you'll need.

Metal and plastic versions are available. Some have battery compartments built in. You'll find internal dividers, removable end plates, feet, punch-outs, wiring channels, and more.

If you intend to paint the finished instrument, metal may be better than plastic. If you plan to cut a hole for a speaker, consider this in advance and have your drilling needs and size requirements in mind. The same goes for overall size of battery holder, circuit, and all controls. The more room you give yourself, the easier the job will be. Following are some examples of what happens when I go looking for ways to house or rehouse a circuit in a project box.

Photon Clarinets

The photon clarinet is a light-sensitive instrument played by hand shadows. Being one of my original designs, it always needs a case of some type to house the electronics.

Pictured are three versions of this circuit, one housed in a rectangular, metal project box, another in a slim, plastic housing, and another in a curved, plastic case (see Figures 11-7, 11-8, and 11-9).

FIGURE 11-7: Photon clarinet in metal project box

FIGURE 11-8: Photon clarinet in plastic project box

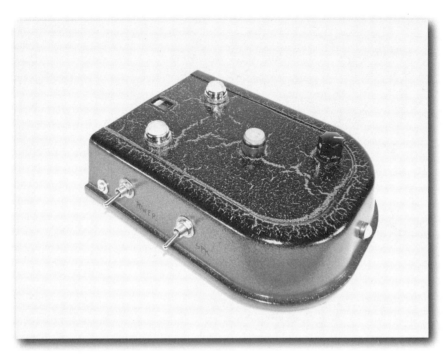

FIGURE 11-9: Photon clarinet in curved plastic project box

Thinking Outside the Box

All kinds of things can be used as project boxes. Just browse the local thrift shop with circuit housings in mind. All kinds of things will pop out at you.

Again, in the following examples I'm still thinking along the standards of compact design and wise use of space, just as I would if resorting to the manufactured project box. But now I'm thinking outside that box.

The Sub-Chant Generator

Another original circuit, this time a stereo human-voice synthesizer. The housing is an old wooden silverware case. Metal panels surround the keyboard, drilled and cut to accept controls and speakers. I added wooden ribs inside the wooden box for the steel panels to locate against and fasten to (see Figure 11-10).

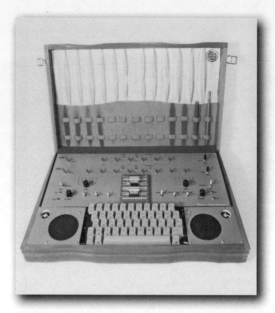

FIGURE 11-10: The Sub-Chant Generator in wooden silverware case

The Harmonic Window

This metal housing is that of an old "AnsaPhone," a precassette telephone accessory for recording incoming calls. After being gutted, it makes a wonderful instrument housing providing a built-in speaker as well as front-door access to my battery compartment. The circuit within this heavy metal housing is a very bent sampler from inside a teddy bear (see Figure 11-11).

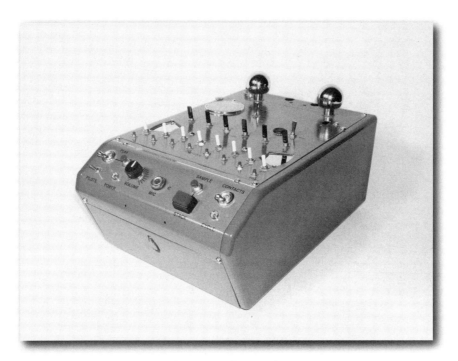

FIGURE **11-11: The Harmonic Window in metal "AnsaPhone" case**

Reverb Horn

I was working with a reverb/amp circuit combo and wanted to turn the sliding potentiometers of the reverb circuit into drawbars, as on the classic Hammond organs. Because a slide pot's knob travels in a linear motion straight back and forth, I knew that a simple solution was near.

Realizing that I just needed to attach an arm to each of the slide pot's shafts and provide a guide for each arm to keep it on track, I browsed the thrift shops with this in mind. The housing also needed to hold the battery compartment, the amp, and a small speaker firing straight up through a hole and into an antique wooden radio horn.

What you're looking at beneath the horn is a large wooden salad bowl turned upside-down, bought for 50 cents at the local Goodwill store (see Figure 11-12).

FIGURE 11-12: The Vox Insecta sound system

The holes drilled in the front of the bowl allow the drawbars to project through, as well as provide a guide to keep their motion level and parallel to each other.

All the needed internal components were arranged around the inside of the inverted wooden bowl (see Figure 11-13).

Rubber feet raise the bowl off the ground and improve the bass response of the speaker.

For a pilot light, I used an ultra-tiny "grain of wheat" bulb. This is a tungsten bulb rather than an LED. I dipped the bulb in transparent violet glaze and installed it in the nearly invisible $\frac{1}{32}$" hole above the drawbars fashioned of brass rod stock and tiny brass drawer pulls. The power switch activates circuit and light, the old-fashioned tungsten glow working much better than an LED in this other-era configuration.

FIGURE 11-13: Amp and digital reverb inside the horn base

Art for Art's Sake

In all the previous examples, wise and reasonable decisions were behind space usage and configurations. There are, however, other potential housings that cannot be ignored. Either their lines are asking to be extended into control configurations or they're already sculpturally on-theme with the project in mind. So, wasted space is not wasted at all.

The Video Octavox

This time the housing is that of an old telephone handset amplifier. Audio oscillators are played by photo cells on the ends of the coiled cords when the cords are suction-cupped to a video screen. A footlighted model of the sun rests in a hole where the earpiece of the telephone receiver was meant to rest. Another original, true-theory circuit, built from scratch (see Figure 11-14).

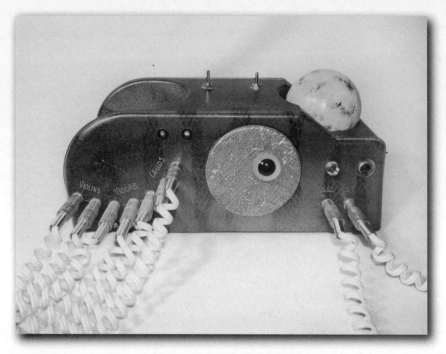

FIGURE 11-14: The Video Octavox in unusual telephone handset case

The Vox Insecta

Here's yet another original design, and this time it's a perfect example of the "build it and bend it" aesthetic (see Figure 11-15).

FIGURE **11-15: Vox Insecta built into 1940s stenograph case**

Here, a simple and theory-true "tone burst" audio amplifier was designed. When it was up and working, I bent the circuit as usual, this time through various capacitors to alter tone and timing.

Finding loads of great bending paths, I needed to assemble a multiple-switch matrix to access them all. I remembered that a stenograph (stenographer's typewriter for writing shorthand in court rooms) had a row of printing blocks that were all arranged in a straight line. Could these work as switches somehow?

It turns out that the spacing of the print blocks was exactly the same as the spacing of the printed circuit traces on Radio Shack's "Experimenter Printed Circuit Board," part number 276-170.

Soldering sections of thin brass rod stock to the cut-down board and then positioning it behind the print blocks allows the rods to contact the board when the stenograph's keys are pressed (see Figure 11-16).

FIGURE 11-16: The Vox Insecta's custom switchboard

The stenograph is now an electronic keyboard instrument.

But I could have done the same with a row of pushbutton switches, right? What really convinced me to use the stenograph was its shape. The circuit in question was an insect sound synthesizer, and the 1940s stenograph's lines were already insect-like, like a stylized beetle or cicada. Adding antique glass telephone pole reflectors as eyes, backlit by fluttering orange LEDs, gives life to the iridescent green body.

Insectaphone

Recognize the housing here? This is an aquarium light that always turns up in the secondhand shops. Liking its odd trapezoidal shape, I designed another, somewhat stripped-down version of the Vox Insecta into it. The voice activation lever extending from the hole at the left is another brass rod, this time an extension of the feeler arm of a micro-switch deeper inside the case (see Figure 11-17).

FIGURE 11-17: Insectaphone in aquarium light case

The Dworkian Register

Named after John Dwork (Frisbee champion, Grateful Dead author, and Grand Weevil of the Phurst Church of Phun), the Dworkian Register would be a great waste of space were it not for the fact that, well, it isn't—even though it's pretty much empty. Here the housing is used to convey style. The actual electronics could fit in a fraction of the space (see Figure 11-18).

Johnny, an early circuit-bent performer, incorporates unusual/interesting/ridiculous sounds into live work—such sounds as those provided by the noise and music strips found mounted alongside the pages of children's electronic talking books. The Dworkian Register bends these sounds way out of context and in four-voice polyphony.

Its bulk, that of an old, mechanical calculator, merely supports the sound circuit strips mounted on top. But the impression is that much is happening in the vacuum within its hollows, various lights flashing, as it sits there rich with mysterious dials, surrounding you with unidentifiable, four-channel surround sound. But, yeah, it's mostly empty.

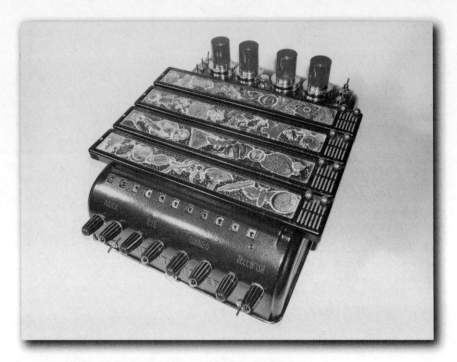

FIGURE **11-18:** Dworkian Register in metal adding machine case

Photon Clarinets

I presented some photon clarinets a little while back in the project box section. There, I tried to match housing with needed size to keep things concise and simple. Here's the other side of the equation.

The monolithic housing is that of an electronic ice crusher, c. 1950–1960s (see Figure 11-19).

Now the speaker rests in the rectangular aperture where ice was fed to the grinder blades for July lemonade on the patio. The ice drawer reveals the battery compartment, and glass-domed sensors plug into the pink-champagne-sparkle sides.

And what do you think the case looking like a smiley face is? It's a floor polisher, minus gears, pads, and all the rest (see Figure 11-20).

All that remains is the rotor housing, recut for oversized photo cells, some pots, and switches.

FIGURE 11-19: Photon clarinet in antique ice crusher case

FIGURE 11-20: Photon clarinet in floor polisher case

Voice Toys and Mr. Coffee

Here's a really cool housing application. As you shop the charity outlets for circuits to bend, you're bound to run into the small voice-changing megaphones meant to make kids sound like robots or parrots, or be just plain louder (what were they thinking?). These are often bendable and in need of a larger housing for added controls.

Just waiting for the transformation are any number of two-part water carafes. Some, like the Mr. Coffee carafe shown, are designed to filter water from the top half to the bottom. Building a megaphone circuit into these is usually easy. Mount the electronics in the top with the speaker firing out the bottom of the filter chamber (just cut out the bottom of the actual carafe with a hole saw). Hold the pouring handle instead as you would the handle of a megaphone, and charm the karaoke crowd (see Figure 11-21).

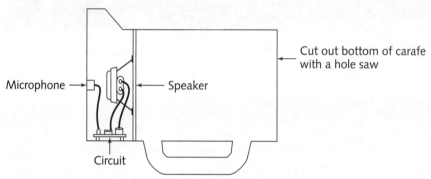

FIGURE 11-21: Water-filtering carafe used as megaphone housing

Installing Speakers

If you've chosen an unusual housing, you'll probably need to create a port of some kind for your speaker to sound through. Let's look at a few possibilities.

If the speaker is really small (up to 1.5 inches), you can enlarge a hole using your hand bore set. Bring the hole up to an inch or so across.

Next, glue a grill cloth or fine metal screen to the speaker and trim off any excess (see Figure 11-22).

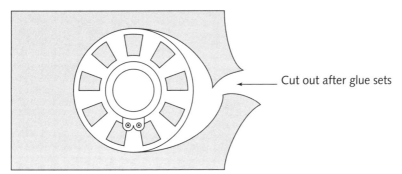

Cut out after glue sets

FIGURE 11-22: Trim cloth glued to speaker rim

Finally, glue the speaker to the backside of the hole behind which it is to reside. Bolting is also an option if you want to drill mounting holes around the speaker's edge.

If the speaker is larger, the best bet is to use a hole saw mounted in either a hand drill or, better, a drill press. Hole saws are sold in sets as well as individually.

Steer away from the sets that rely upon a central spindle upon which the varied blades fit. The blades in these sets are flexible as well as ill fitting, both problem points. Obtain instead one of the more expensive, nonflexing hole saws—the kind that mounts on a hole saw drill bit, or "mandril."

After the hole is drilled, you'll be at the grill cloth stage again, unless you want to find a cover that mounts to the outside of the case like an automotive speaker grill. Drill mounting holes and use bolts instead of glue for larger speakers.

The Speaker Cabinet Project Box

Easier still is using a box with a speaker already built in. If you shop surplus, you'll undoubtedly run into various speaker enclosures from intercoms, computer sound systems, and the like. Many of these will serve as really great housings for circuit projects and save you the hassle of using a hole saw to mount a speaker in a standard project box.

Additionally, many older extension speakers and intercom speakers have great design lines and look so much better than the bland plastic and aluminum project boxes you might be tempted to use otherwise. Here's another photon clarinet, this time built into a 1940s intercom salvaged from an old garage (see Figure 11-23).

FIGURE 11-23: Photon clarinet in antique intercom case

Creating Templates for Reproducing an Instrument

Getting a good control layout can take a while, especially if you need to position your added components precisely to maintain important clearances. If you intend to reproduce the instrument for yourself, friends, or for sales, making a template for marking component locations is a very, very good idea.

All you'll need for this is transparent film sheets (such as the 3M overhead transparency film, part # 5470296) and your ultra-fine felt tipped Sharpie brand, permanent ink marker. Here's how:

1. After you have the instrument marked for the new control locations, and before you drill even the pilot holes, lay a sheet of film over the instrument.

2. Either position the film so that its edges line up with edges on the instrument (see Figure 11-24) or position the film at key detail points of the instrument's design (see Figure 11-25).

These will be used to align the film with the same features on the next instrument.

Film located at edges

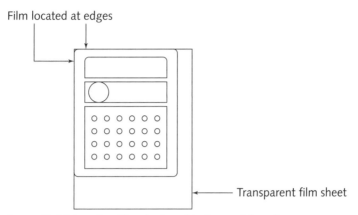

Transparent film sheet

FIGURE 11-24: Locating film at edge of instrument for reference

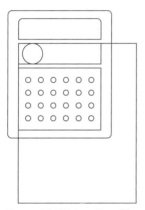

FIGURE 11-25: Locating film at
detail of instrument for reference

3. After you've decided how to position or mark the film so that you can locate it the same way next time, make a dot on the film with the pen corresponding with each dot on the case where a hole will be drilled and a component will go.

4. After you've dotted all positions on the film, use a compass point, a sharp awl, or any other pointy object to poke a *small* hole in the center of each layout dot on the film.

5. Position this perforated film over the next instrument, align with sides or details, and, using the pen again, mark a dot through the tiny holes and onto the new instrument.

With care to position the template precisely, and with careful marking all through the process, this procedure will streamline series production. I often design several templates for a single instrument. Separate templates might be needed for different sides of the instrument, or for difficult, indented, or oddly shaped areas.

Bending Toward Concept: Two Examples

Although you clearly can't attempt to make a bent circuit meet your needs the way a store-bought synthesizer is designed to, you're not completely at its mercy, either. You can, to a degree, have it serve your will. Two territories will stand as examples of this: bent designing toward the concepts of instruments played by the environment; and instruments as ambiance synthesizers.

Environment-Played Instruments

As you bend instruments, you'll find many circuit areas that will present great variable output when run through potentiometers. And potentiometers are, you'll recall, variable resistors. If you use photo cells in place of potentiometers, you instantly have an instrument that will respond to changes in lighting.

As mentioned earlier, you can substitute solar cells for batteries to the tune of three cells per 1.5-volt battery you're replacing. Wire them like batteries, in "series," all soldered in sequence, positive to negative on down the line (see Chapter 9, Figure 9-26).

Now, in addition to the photo cell modulation, the overall power of the solar cells will vary with the brightness of sunlight, causing the instrument to sound very different from morning to dusk as its supply voltage changes. My solar bug box series is a good example of this solar-powered, solar-modulated instrument technique (see Figure 11-26).

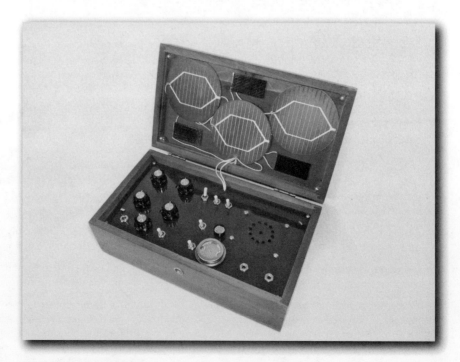

FIGURE 11-26: Solar bug box

But that's not all. Available is a plethora of other sensors that can be soldered into the middle of your bent circuit paths. These include temperature, humidity, water level, water direction, water velocity, wind direction, wind velocity, barometric pressure, UV radiation, ground-moisture sensors, and more. There is a vast territory to explore! Search surplus outlets for sensors.

Ambiance Synthesizers

We think of instruments for creating music to be the things up there on the stage, the guitars and drums we've shelled out money to hear. But for the greater history of human experience, music was played outdoors in the midst of environmental sounds. There is a balance missing when we bring music indoors, into the sonically sterile environments of the recording studio or acoustically deadened recital hall.

Of course, we do this for a reason: "good" audio listening and reproduction. Understood. I have a sound studio, too. But environmental sonic ambiance can be pretty interesting.

Toward this goal I built the Dworkian Register, the surround-sound instrument I mentioned a little while ago (see Figure 11-27).

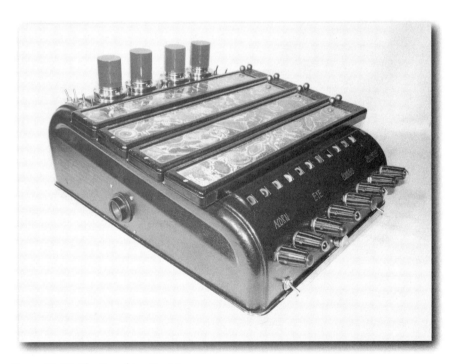

FIGURE 11-27: The Dworkian Register

The cartoon sounds of the book strips become all new entities when sped way up or slowed way down, exactly the final result of the bending. But the discrete four channels are not meant to be brought out to the stage speakers, but rather to speakers hidden in the corners of the auditorium.

Now a surreal surround sound can be created in which the other instruments, the heart of the performance, can take center stage. The audience may forget the surround-sound elements. But just as the birds of a nearby field subliminally impressed listeners of open air Renaissance music, the Dworkian Register provides similar, though rather fanciful, atmosphere.

Many of the instruments we've just looked at benefit from dramatic finishing elements. Check out this book's companion Web site (www.wiley.com/go/extremetech) to see some of these instruments in full color. And in the next chapter, discover for yourself how the final transformations are done.

Finishing Techniques

A s a hobbyist entomologist (bug kisser), I'd never heard of a singing moth. Moths are silent, or at least without voice. But at one of my northern camps I'd tracked the sound to the stubby weed at the edge of the pond, a small, fluttering plant full of bright green leaves and harboring a large gray moth. Now, this close, I could even see the moth's wings vibrate as the liquidy trill emerged.

My approach finally alarmed the moth, which flashed for the first time its underwings, a bright pair of yellow-and-black eyespots meant to frighten predators like me. The trill, raspier and shrill, accompanied this burst of color as its owner, a well-camouflaged green tree frog, leaped past the moth and into the water.

No, brightly colored moths do not sing. They're not designed to. That is, unless you're the designer.

Various Painting and Surfacing Techniques

As I develop an instrument, as I listen to the new bent voices and consider the instrument's lines, a personality begins to emerge. To take this personality further by means of visual applications seems natural. It's a challenge. But it's also a joy to see an instrument develop like this.

Before any control ornamentation such as special body-contacts or LED housings can be added, before mounting switches or doing anything further at all, painting should be considered. Lots can go wrong while painting. But at the same time, careful usage and even rule bending can yield simply wonderful results.

Paint Choices

Whether you're using an airbrush, experimenting with marbling or other faux finishes, or even using canned paints and brushes, painting techniques are many—even endless. I hope you'll explore these and more. But for right now, I discuss spray paints because everyone has them, they're cheap, and if applied creatively they can be very effective.

As of this writing a series of special spray paints for plastics has been introduced by Krylon. The series is named "Fusion." My first tests have been successful. Right now I have a Fusion example out in the weather and all looks good.

But I've also been using the standard spray paints for decades with satisfactory results, even on plastics. No, results are not even. Plastics vary greatly in composition. But a good first wash in soapy water followed by a thorough warm-water rinsing is usually effective—standard spray paints seem to adhere thereafter reasonably well. Let's look at what's available.

Flat Finish

Flat finishes give rich color with little or no surface shine. They're usually fast drying (often faster than gloss coats) and are marginally more forgiving then gloss paints if accidentally sprayed on a little too thickly; they won't run or drip quite as quickly.

Flat coats are better than glossy for an undercoating or primer. Flat paints scratch easier than glossy. Always look for "fast dry" on the label.

Gloss Finish

Glossy paints are the types used most often for the final instrument finishes. When dried well and fully cured, high gloss paints are scratch resistant, color fast, and reasonably durable. Again, look for "fast dry" types.

Acrylic Latex Sprays

New to the market, latex spray paints do have their applications. Their low-fume formula is certainly attractive. However, their longer dry times, thicker build-up, and spotty performance in adhering to plastics keeps them from being my first choice.

Specialty Paints

We're lucky to have so many specialty spray paints these days. When I began to paint instruments back in the 1960s I was limited to only a few colors and even fewer metallics. Fluorescents were unheard of, and all kinds took hours and hours to dry. Times have changed!

Metallic

Metallic paints contain fine metal dusts and are usually high gloss in finish. Their metallic reflections are the classic stuff of hot rods and motorcycles. Metallic paints look great on the curves of instruments and are nice for accenting an instrument's molded lines.

Hammer Tone

Hammer tone is often another metallic paint, though usually not as glittery. The unique quality of hammer tone is that it settles in little depressions as it hardens, thereby simulating the look of hammered metal.

Glitter Paints

As do metallic paints, glitter paints contain highly reflective, colored metal particles. However, the particles in glitter paint are usually larger than those in metallic paint and, whereas metallic paint is most often based on a pigmented lacquer, sparkle paint's lacquer is usually without color.

Neon Glow

"Clear Neon" (the product name) is a water-based spray paint that's invisible unless seen under UV. If sprayed onto a flat-white undercoat and then sealed with your usual clear topcoat, you'll have an instrument with a bipolar personality. Demure in daylight, but a party animal when the black lights come out.

Crackle

Crackle is actually two paints—an undercoat and an overcoat. In action, the overcoat shrinks and cracks apart as it dries, thereby allowing the undercoat to show through.

This cracking is based in part upon the fact that the undercoat is a glossy paint. Its slick glossy surface allows the (usually flat) overcoat to slide as it contracts, aiding in the cracking. Be sure to let the undercoat dry thoroughly before applying the top crackle coat.

Here's an experiment. Grab a gloss paint and a flat paint, anything you have on hand. Find a scrap piece of plastic and do the crackle thing. Gloss coat first. Dry. Then apply the flat coat. See whether you get any crackle. Many gloss/flat paint applications will crackle or react in interesting ways (you can speed the crackling with a heat gun).

Stone and Other Textured Finishes

"Stone" finishes depend upon thicker, flat-finish spray paint followed by another coat of spray paint. The second coating contains flecks of material simulating the grain of stone.

Other textured paints are available, too. If this is your thing, be sure to do a practice spray area on the inside of the instrument, in a hidden spot, to check on the paint's performance. *Looks good? Adheres well? Be critical.*

More Multistep Paints

A well-stocked paint or craft store will offer many additional two-step spray finishes. Pearlescent, opal, and bicolor iridescents are now common. As usual, test on an inconspicuous spot before committing your instrument to a ten-year peel.

Pattern Masking

Pattern masking can be as simple as dropping leaves on the instrument before spraying, or as complicated as precisely taped lines laid out in elegant, swirling shapes. Scraps of netting as mask can give the impression of reptilian scales (see Figure 12-1), just as torn strips of paper could simulate zebra stripes. Much to explore here!

FIGURE 12-1: The Video Octavox (For a more-detailed view in color, see this book's companion Web site, www.wiley.com/go/extremetech.)

Powders and Dusts

Although much color shading can be done in the traditional way by using two different-colored paints sprayed from different angles, colored powders can also be used for highlighting or hue blending. Additionally, iridescent dusts of many hues are available for craft work and custom paint mixing.

Micro sparkle is also now on the market. This is a laser-etched particulate nearly as fine as dust, refracting light into the entire spectrum. All these powders and dusts can be added to your instrument's finish, as outlined later.

Pigments

Dry pigments in powdered form are available from art supply outlets. These are available in all colors in both flat and metallic varieties. I keep a variety on hand for various art projects.

Sparkle Dusts

These laser-etched micro flakes are available at craft stores. I like the smallest flakes. All come with a shaker top, a handy dispersal mechanism. As mentioned, some are tinted a specific shade so that you can go with a color complementary to your paint job, a nice effect in holographic sparkle.

Marking Control Titles

Wait a minute—just *what was* that button supposed to do?! After you've painted the instrument, along with the scratched-up gray plastic also went all the original button titles. The Power button looks just like the Record button now. That's inconvenient.

Control titles, or "legends," can be marked on the new paint job in one of two main ways: using pen and ink or rub-on transfers. Each works just fine. The difference is in the look.

Transfers

Transfers are available in differing fonts and sizes. Those made especially for electronics include "roses" (radial degree graphics) for dial markers and are usually a little stronger than those made for graphic artists. They also include ready-made words common to electronics panel legending: ON, OFF, BEND, er, strike that last. But both styles will work for control legends (see Figure 12-2).

FIGURE 12-2: Transfer titles

The trick with transfers is in applying the final sealant—a clear gloss topcoat. Just remember that too much wet gloss can float the transfers right off the panel you've rubbed them onto. Several even, light coats are the answer here, as in all spray painting. More on this in a moment.

Pen and Ink

I like the feel of pen and ink and keep various pens at the workbench, all in easy reach. Many are exchangeable, but not the pens used for panel marking. Why? Because most of the "permanent" pen inks will run like a river when the top sealant is sprayed over them. Really bad news.

There's only one kind of marker I trust not to do this: a fiber-tipped liquid *acrylic* marker. Instead of being solvent-based ink, the acrylic ink will not dissolve in the presence of "thinners," or solvents, the base of most clear sealants. Add to this the ultra-fine tips and all the highly pigmented colors and you've got a real problem solver. Look for the fine-tipped "Painters" markers by Hunt Corp., N.C.

Be sure to let the instrument's paint dry thoroughly. When it's dry, simply mark the control titles as though you were writing a letter (see Figure 12-3, plus find more details in "The Painting Process, Start to Finish," later in this chapter).

Try to buy the entire set of pens and be sure to get the smallest tip size available.

FIGURE 12-3: Hand-inked titles

Final Glosses

Clear spray-on glosses are the final coating to be applied to the instruments. These coats will seal-in the powders or dusts you've applied as well as form a protective topcoat over any control titles you've added to the design.

Look again for fast-drying, high-gloss sealants with "high solids" content if possible (high-solids clear glosses need fewer coats to build up a thick finish).

Designing Your Own Paint Booth

Painting is actually quite easy as long as you're prepared to do a good job. This comes down once again to techniques and tools. Your paint booth can be rather simple and still fit the three main requirements: bright lighting, workpiece accessibility, and fume exhaust. Note that this is a workshop process, unless you don't mind your bedroom changing colors as the weeks roll by.

Lights

Two-point lighting in the form of dual 100- to 150-watt spotlights will work fine. These can be mounted in shades positioned to the right and left of the workpiece, a little above and in front (see Figure 12-4). Clamp-on lights with reflectors work well too.

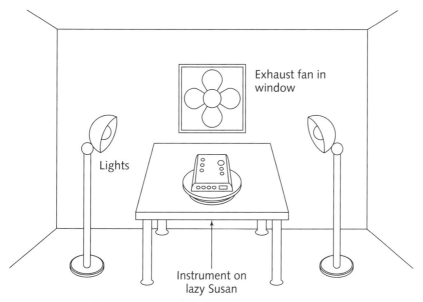

Exhaust fan in window

Lights

Instrument on lazy Susan

FIGURE **12-4: Simple painting setup for venting fumes**

Lazy Susan

Talking about lazy Susans back in Chapter 7 was fun, but putting a lazy Susan to real work is even better. The lazy Susan is the painting platform for your instrument. Being able to rotate the turntable while you paint will make the job much easier than trying to walk around the piece as you go. Position it at a good working height, as in Figure 12-4.

Exhaust

Exhausting the paint fumes out and away is important and is easily done. The simplest and least expensive way to do this is to position the painting station near a window and use a window fan (refer to Figure 12-4).

Be sure that you have good air flow-through (open a door or window on the opposite side of the room), and use old sheets or taped-up newspaper to shield all surrounding areas. The window fan on its highest setting will do a great job of keeping the air at the lazy Susan refreshed, but there will still be overspray to deal with.

If you do a lot of painting, you might want to invest in a filtered booth, homemade or pre-assembled. But for the occasional paint job, the easy setup and teardown of the window fan system can't be beat.

Note that in theory, the electric sparks produced by a motor's brushes are enough to trigger a fire within a volatile environment. In a unventilated room full of dense paint fumes, this might just happen! However, in my experience of venting fumes via a room's crosscurrent through the use of a window fan, I've never experienced such a flashback fire (or "explosion," as the warnings are usually termed). In fact, I've seen frequently operated paint booths designed around a large window fan situated right next to the painting platform (see Appendix B). Nonetheless, my advice is to keep the fan a few feet away from the spray area. If it seems as though fumes are accumulating, seek a better ventilation scheme and take a break to let the air clear.

Want to go deeper? Small, high-velocity, *brushless* fans are available inexpensively through surplus dealers. A matrix of 16 (4 × 4) of these fans situated in an open window will eliminate any concerns over conventional fan electronics. However, no matter what your painting setup is, be sure to *always* keep a fire extinguisher close at hand, both at your paint area and your workbench. And *never* allow paint fumes to grow thick around any potential source of combustion.

Mask Up!

Be sure to *always* wear a mask made for spray paint while painting and until the air completely clears. Many styles are available; different manufacturers use different materials. I prefer the dual-respirator style. Just be sure that the mask is rated for spray paint and that you change filters as recommended in the mask's instructions.

The Painting Process, from Start to Finish

Here's the road to success for painting, step by step. Tempted to skip a few steps? Don't! All are important for a good job.

1. Prepare your drilled-out instrument case by washing the plastic to be painted with warm, soapy water. Rinse thoroughly and dry completely.

   Either remove or mask with masking tape any sections not to be painted, including power or output/headphone jacks.

2. Prepare the painting area as outlined previously and have all paints shaken and ready nearby.

3. Place the workpiece on small wooden blocks atop the lazy Susan.

   Saw a long, square, $\frac{1}{2}$" diameter wooden dowel into short sections and keep these on hand near your painting area.

4. If the instrument is to go through a vast color change (say, the original plastic is green and you want to paint it red), you'll need a white undercoat to make the new color true and radiant. If you're staying in the same color area, you might not need a white primer coat. For this example I'm assuming that a white coat is needed. Grab your well-shaken can of flat-white spray paint.

5. With lights and fan on, turn the lazy Susan so that one side of the instrument faces you.

6. Now, this is important. Your sweeps of the spray paint can, along with actual spraying, must begin *before* the workpiece and end *after* the workpiece. This will result in overspray at both the beginning and end of each stroke (see Figure 12-5).

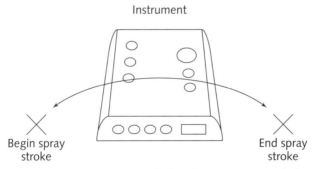

Instrument

Begin spray
stroke

End spray
stroke

FIGURE **12-5: Begin paint spray before the instrument; end after.**

Overspraying like this keeps the painting process smooth and as sputter free as possible (sputtering usually occurs the moment you depress or release the spray head).

7. With the first side of the workpiece facing you, push the spray head of the flat-white paint can and begin to spray paint off to one side of the workpiece, missing it completely. With the paint still spraying, and while you're keeping the spray head about 10 inches away from the plastic, sweep the spray across the workpiece in a smooth line to the other side until you're once again completely missing the target. (Each sweep should take no longer than a second or two—be smooth and somewhat quick.) At the end of the sweep, you can release the spray head and stop the spray.

8. Repeat this spraying technique, back and forth, up and down the instrument in tight parallel rows, incrementally, until the areas facing you have a light, even dusting of paint. You're not going for a solid-white coating yet. *Just a dusting*, a white haze at best.

9. Using the lazy Susan, rotate the workpiece to the next side and once again paint by over-spraying in long, smooth sweeps. Repeat until the first side is facing you again.

10. Now the workpiece should have a light, hazy coat of white paint on it. Not a solid white yet. You probably painted too thickly if it looks at all solid white. And you *certainly* painted too thickly if you caused the paint to run anywhere. You're looking for misty, but even, coverage at the end of this first coat of flat-white primer. Let it dry completely before you proceed to the next step.

11. When the paint is thoroughly dry, repeat the whole process with more white paint. This time, when the first side comes around again you should be getting closer to a solid-looking flat white. Let it dry once again and, if the flat white is not yet solid, repeat the process, applying one more very light coating of paint.

So much for the primer. Now for the fun parts!

1. Grab the well-shaken topcoat color (red, in my example) and repeat exactly the painting process of the previous steps, with overstrokes and very light coats being the key to success here. Again, three light and hazy coats of red paint, without their looking like a solid color until the last is applied, is what you're after. Light coats will eliminate drips and runs, exactly what you're trying to avoid.

2. If you don't intend to add dusts or control legends, skip right ahead to Step 3. But if you do want to add these:

    a. To add pigmented dusts or sparkles to the paint job, turn off the fan and hold a small amount of the dust about two or three feet above the instrument before the last paint coat has completely dried. Sprinkle the dust or sparkle sparingly, sweeping your hand around in the air above the instrument. This should help disperse the material evenly. Remember that pigmented dusts will show up much better on light-colored backgrounds! Sparkles work fine on all colors.

    b. To add control legends (by ink or with transfers), you must let the paint dry completely, which might take a few days, even with "fast dry" paints. Okay. You can push this schedule. But give the paint as much time to dry (and cure) as you can afford. If not, the transfers will distort the paint, and pen nibs will catch and drag on the painted surface, also distorting the paint. Allow the acrylic paint from your pens to dry thoroughly before the next step.

3. If you didn't add dusts or titles to your painted instrument, you don't really need a final sealant coat (even though a clear topcoat can make the main red coat look more vibrant). If you did apply dusts or sparkle, a clear topcoat is mandatory both for looks and its function as a sealant. The clear gloss will increase the sparkle's radiance tenfold and will seal the inks with a nice protective coating.

Apply the clear sealer just as you did the paints, one very thin coat at a time, until a nice and even gloss is achieved. Let it dry very thoroughly before even thinking of handling.

Important Dos and Don'ts

The most important *don't* of spray painting? Don't pause in midstroke. You'll immediately spray too much, and an ugly run will appear. Keep the spray can moving at all times. And resist the temptation to rush things. Applying slow, light, even coats is the key.

And *do* use either a heat gun or a small forced-air space heater to speed drying, especially if you've painted in a high humidity environment. A heat gun can drive the haze out of wet paint that sometimes occurs in clear coats if applied on humid days. As with the spray can, keep the hot air of a heat gun moving all the time. Directing the gun's airflow at a single spot on the newly painted instrument can distort the paint, the plastic, or both.

You can also use a forced-air space heater to speed drying. Position the painted workpiece in front of the heater at a distance from which gentle, warm air blows over the paint (see Figure 12-6).

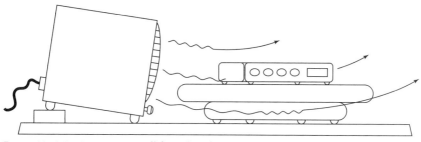

FIGURE 12-6: Positioning a small forced-air heater to speed paint drying

Hot air is not needed. Again, positioning the workpiece too close to the heater can distort paint or deform plastic.

Hardening and Handling

As mentioned, fast-drying paints dry to the touch rather quickly. But for them to fully harden and cure takes much longer, usually days. Be sure to let this time elapse before handling the workpiece or you'll end up with fingerprints on the finish, and the paints will bulge where you tighten the component hardware.

When the paint is fully dried (test paint for hardness with a fingernail in a spot that will be hidden by a component, such as the edge of an empty switch hole), you can proceed to mount your components and begin the soldering process. However, all post-painting bench work must now be done on a folded pad of deep-nap terry cloth (or another firm but very soft surface). You've put a lot of time and effort in the finish. A soft work surface will go a long way toward protecting the paint job.

Fixing Those Painting Mistakes

Drips and sags occur when paint is sprayed on too thickly. Fixing these mishaps is no fun at all. The best solution here is to be very cautious with your painting and always err on the side of too little rather than too much paint. Still, drips happen. There are a couple ways to get around this mess.

Sand It Out

Let the drip dry, sand it out, and start that coat again. I know. No one wants to do that. But that's how it's done.

Use a medium-grit sandpaper and be prepared to toss it out as you go, because the paper will quickly become clogged with paint. Last, use a fine-grit sandpaper to smooth the bad spot as well as possible.

Hide It!

Sneaky, but hiding a drip or other small paint flaw is often a good alternative, especially if the flaw is on a final (and beautiful) coat and you don't want to have to paint the entire instrument again.

What would be large enough to hide the flaw? Can you mount another switch there? A body-contact? A pilot lamp? A pot? How about a nameplate? Even a dummy bolt with a chrome head looks better than a paint drip. Only you will know it holds in place nothing more than your instrument's dignity.

Not done yet. Whenever there are enough of any unwanted things in the world, no matter what they are, someone always seems to come to the rescue. There are so many unwanted holes, it turns out, that industry has stepped in to, well, fill them all. Or at least as many as possible.

Look for these snap-in, chrome or plastic mini hubcaps in the nuts-and-bolts section of your major hardware store. You'll probably have to widen the hole a bit with your hand bore to get the next-size-up hole cap to fit.

Adding Control Ornamentation

This is a fantastic area to explore if you enjoy "theater" or further extending an instrument's personality visually. I collect all kinds of unusual items to incorporate into instruments. Wire in all colors and sizes, antique pilot lamps and knobs, vintage switches and other components from electronics' classic era of deco style with glass, brass, and colored Bakelite trappings (see Figure 12-7).

FIGURE 12-7: Odd component collection, including a radioactive doorbell button, at upper right

But also I incorporate many nonelectronics-industry pieces into instruments, such as prosthetic human glass eyes and antique German Bauhaus–era black glass buttons. I collect *many* antique buttons, especially stained shell varieties (see Figure 12-8).

Read on.

FIGURE **12-8:** Vintage clothing buttons for transforming
plastic buttons on instruments

Body-Contact Variations

In the project section, you'll be using commonly available turned brass balls with a hole about halfway through (blind hole), threaded for a small bolt and available in lamp supply sections of well-stocked hardware stores, or online (www.customlamp.com, for example). Although these are efficient, there are more interesting ways to go.

First stop is the drawer-pull (knob) section of the "home improvement" store (we all improve our homes by spending too much for simple hardware, don't we?). Metal knobs of all kinds can be found, most of which are ready for body-contact usage. There are pulls shaped like twigs, shells, insects, leaves, and far more. Of course there are the more usual knobs—orbs, flat discs, and "mushroom" shapes in all sizes.

They all fasten to your instrument with a bolt or two. You can thread the bolt through the eyelet of an electrical terminal before threading it into the knob, thereby providing a ready solder point. Or just solder the eyelet to your body-contact wire first and then thread the knob bolt through (see Figure 12-9).

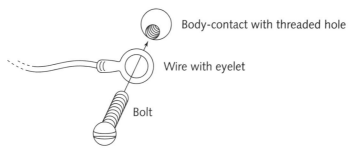

Body-contact with threaded hole

Wire with eyelet

Bolt

FIGURE 12-9: Wire–to–body-contact

When you buy your knobs, be sure to also buy shorter bolts (such as $\frac{1}{2}$" long) because the supplied bolts will be too long. They were meant to pass through a door, after all.

If you have a bolt cutter (a pliers-like device), you can cut the supplied bolts down to size. Be sure to thread a nut onto the bolt before cutting so that you can twist it off over the cut, reshaping the bent threads there (see Figure 12-10).

Nut   Cut

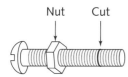

FIGURE 12-10: Put a nut on the bolt you're cutting and unscrew after you're done.

This is a good idea even if you're using a hacksaw to cut bolts because threads are always damaged.

Using knobs as body-contacts is not as straightforward as it may appear. Some knobs are coated with an antitarnish coating. Usually these are brass, bronze, pewter, or copper looking. The coating does not conduct electricity well at all, making these useless for circuit-bending. On the other hand, bright chrome is almost always uncoated. The rule of thumb here: If the contacts don't seem to be sensitive, they might be clear coated. Try a bright chrome style.

Looking at the Species Device, a sample-based body-contact instrument, you'll see drawer pulls as body-contacts (see Figure 12-11).

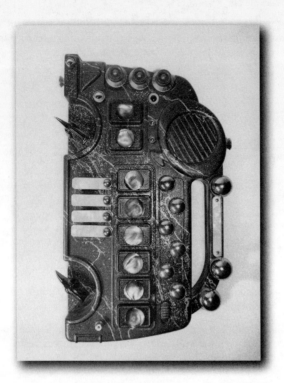

FIGURE **12-11: The Species Device's many body-contacts**

Every knob is wired to the circuit, all 13 of those you see.

Simpler still are decorative bolts. These are bolts whose heads are not hex, Phillips, or slotted. Most common at the hardware store are "carriage bolts" with plain, domed heads. But there are many decorative bolts available, especially in surplus, in head sizes from ¼" to several inches across. Again, simply insert the threaded shaft into the hole and fasten with a nut from behind with your electrical eyelet slipped on first.

Hand tools are available for "tapping" holes—milling out threads in the sides of a hole you've previously drilled so that a bolt can then be threaded into the hole. This way, any drillable metal object can be used as a body-contact. Certain abstract jewelry, for example, will lend itself to this technique.

Finally, metal plates can be used as body-contacts. Because getting a plate hot enough to solder to is difficult (it's possible, but often there's too much metal to heat up enough for a good solder joint to be made), you can affix the plate to the instrument with small bolts, such as a guitar pick guard is affixed, and use one of these bolts to hold the electrical eyelet as before (see Figure 12-12).

Again, plates of differing types are available in the knob section of the well-supplied hardware store.

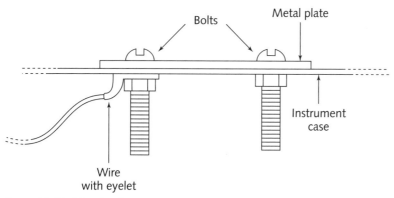

FIGURE **12-12: Wire-to-plate connection**

Switch and Button Transformation

I've already given an example of how the keypad of a stenograph can be transformed into an electronic switching system (Chapter 11, Figure 11-16). Keep watch for other mechanical mechanisms that, with alteration, might serve well as electrical switch actuators. But what I'm describing now is transforming the switches and buttons themselves.

Using the example of the Species Device again (see Figure 12-11), old piano key "ivories" (actually celluloid) were reshaped and fitted to the instrument case just above the four circular pushbuttons of the toy's original design. (These buttons originally looked like the wheels of the train the animals are riding on.) Pressing the piano keys now also presses the buttons hidden underneath.

Also on the Species Device, I've fastened mother-of-pearl buttons to the remaining animal voice keys. This transformation of plastic train wheels and animal faces into piano keys and oversized "accordion" buttons pushes the instrument way into another reality.

Extending the feeler levers of micro switches (see Figure 12-13) allows you to design countless actuator buttons. As shown earlier, in one of my Insectaphones (Chapter 11, Figure 11-17), I control a voice actuation circuit with the extended arm of a micro switch (see Figure 12-14).

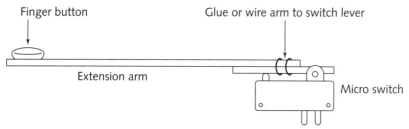

FIGURE **12-13: Extending the feeler lever on a micro switch**

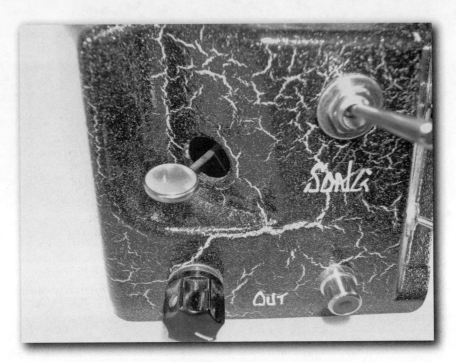

FIGURE 12-14: Micro switch arm extending out of instrument's side

In this example I've tipped the extension with a small, gold-rimmed button. On the same instrument I've also topped the stock pushbuttons with identical gold-rimmed buttons to tie all switches together visually.

LED Transformation

Although an LED mounted in a precise hole looks fine all by itself, transforming this light source into more interesting configurations is easy. The most obvious of these is simply installing the LED into a pilot housing of one type or another. If you shop the surplus marketplace, you'll find many variations, including fine examples of pilot lamp housings with mechanical shutters (to adjust brightness), finely cut stained glass, and seemingly endless abstruse titling (see Figure 12-15).

Many other light-transmitting items can be used for LED housings or lenses. In the example of the Dworkian Register, I've used four vintage Lucite "light pipes" at the rear to cast ultra-bright red envelope LED light into the room (see Figure 12-16).

FIGURE **12-15:** Assorted pilot lamp housings

FIGURE **12-16:** Light pipes on the Dworkian Register

These light pipes, probably from the 1950s, are pink Lucite rods meant to be screwed onto the end of a flashlight for use as a signaling device. Marbles could be used in the same way.

Glass eyes are great for LED lenses. These can be rear mounted in any area with enough space for the eye and LED combination (see Figure 12-17).

FIGURE 12-17: Glass human eye on bent guitar

Human glass eyes made for the medical industry are superb transmitters of LED light. I'll at times use the same eye as both steady power lamp and throbbing envelope lamp, shining two LEDs into the same eye from behind (see Figure 12-18).

Unlike the glass animal eyes made for taxidermy and puppeteers, glass human eyes are hard to find, as well as expensive. However, they disperse the light within them wonderfully.

I've also used antique mechanical doll eyes ("sleepy" eyes) as LED housings by drilling the back of the eye's housing so that the light of an LED can shine in. These eyes open as the instrument is brought to an upright position, and they light as both power indicator and envelope lamp on the Species Device (see Figure 12-19).

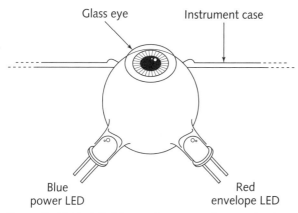

FIGURE 12-18: Back lighting a glass eye with LEDs glued in place

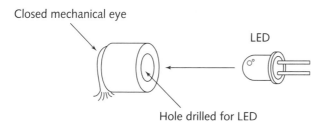

FIGURE 12-19: Illuminating a "sleepy" doll eye

Eighteen Projects
for Creating
Your Own Alien
Orchestra

part

IV

When you have one instrument before you, you begin to think as a composer. But with an orchestra in front of you, you're also an arranger, suddenly capable of taking your musical thoughts much, much further.

The Instrumentarium: An Overview

You're about to dive head first into an alien world! You'll be constructing chance music machines and new language generators, designing body-contact instruments, and way more.

For each project, you'll have construction broken down into sensible, step-by-step procedures. Included for each project are guides for opening the instrument's case, finding bend points, taking notes, choosing a control layout, determining specific case considerations or alterations, drilling holes, mounting your controls, and choosing soldering techniques. The guides also help you with circuit testing, reassembly, musicality, and operating instructions.

Often, I just touch upon many of these subjects because the diagrams allow construction to proceed without any reading at all. That's fine, especially if you're already comfortable with electronics. In a few cases, however, special considerations are mentioned. Giving the construction text a glance might save you a little trouble.

Tip

Be sure to check the companion Web site (www.wiley.com/go/extremetech) for high-resolution pictures of the project circuits and their bending points.

Look for the target circuits at all the secondhand and charity outlets in your area. Goodwill, Salvation Army, St. Vincent DePaul, and AMVETS are my main outlets here in the Midwestern U.S.A. As mentioned earlier, look also at flea markets, garage sales, and surplus sales, as well as at old stock on hand at retail outlets. These circuits also show up in the classified section of newspapers, in local trade-or-sell publications, and on eBay.

All bends outlined herein are original to my own explorations. All were found within the clear illogic of circuit-bending using the chance techniques described earlier in this book. Although the bends you'll be doing are the ones I use and know to be dependable, they are certainly not the only bends possible within these circuits.

In fact, I have many additional bends in my notes. If you want to explore further later on as you find duplicate circuits, by all means! Such further exploration is what the art of bending is all about.

So, why haven't I included everything in my sketchpads? Simply stated, the following projects are from the notes I trust the most (besides that, there's just not enough room for my seven-notebook stack). I've left things out that I know are workable on some circuits but disastrous on variants of the same circuit. I've left out some things that I know lead to mechanical trouble. I've had to make the needed compromises to secure your success and your circuit's longevity.

In striking this balance, I look chiefly to the emerging personality of the bent instrument and implement bends that push this personality forward. If space is limited, I'll often choose compositional bends over tonal bends because tone changes can be created-in later on in the studio using traditional gear. In each of the following projects, my focus has been on giving you, as safely as possible, the biggest bang for the bucks.

Know as you proceed that, even sticking to the plans in the following section, you can still run into trouble. Almost always the trouble is due to variant circuits produced via different manufacturing runs over a single product's lifetime. A simple redesign of a circuit can result in a burnout, even when the circuit is bent exactly as before. Too, well-used circuits can be on the verge of natural death anyway, prior to the pressures of circuit-bending. Don't let the occasional burnout halt your journey.

Be confident in yourself as a bender. Your work here will result in dependable instruments as long as you follow the construction techniques, assembly plans, and guidelines for use.

In most cases, you'll be working from reproductions of actual circuit boards. In some examples, you'll base your work on simplified drawings of circuit components. Either way, you'll have all the information you need to succeed.

Let the following projects introduce you to the field. You can then go further yourself, allowing the threshold of invention to help you discover your own fascinating, and totally *original*, instruments. Who knows what you'll find or hear!

Project Conventions

The following projects look to a modest set of conventions in the way circuits are portrayed. These conventions are divided into three categories: switch depiction, LED orientation, and the circuit standards covered in Appendix A that you'll be referred to.

- **Switch depiction:** All toggle switches are depicted as single-pole, double-throw (SPDT) style. That is, it's a toggle switch that has three poles (soldering lugs) on the back and two "on" positions. In reality, all toggles shown with three lugs but with just two wires attached could as well be single-pole, single throw (SPST) switches—that is, a toggle switch that has only two lugs and one "on" position. In fact, if you're buying one switch at a time, you'll save a little money purchasing SPSTs instead of SPDTs. So why do I depict SPDTs? Simply because these are the switches you'll most likely find cheaply in quantity, in surplus. Because they can also be used as the simpler, two-lug SPST (the application you'll see in most of these projects), I've shown them wired as such. In addition to this, I want to be clear on which two lugs are used when wiring a SPDT as a SPST, a very common application in electronics (see Chapter 6 for more on switch use). But by all means, feel free to buy SPST toggle switches instead of SPDTs for any switch depicted that, again, gets only two wires soldered to it.

Note Either a two-lug or three-lug toggle switch will work fine for the projects in this book.

- **LED orientation:** LEDs are usually designed with a different length for each lead. At one time, in the Library of Alexandria, there was a papyrus scroll detailing exactly what LED lead length meant. Some scholars argue that the longer lead must be positive and therefore the anode because bigger is always better and therefore . . . positive. Historians confound this conclusion with their own. The fire that burned Alexandria was huge: Bigger is not always better. The longer lead must then be the negative lead, or the cathode. In response to this lost knowledge and subsequent confusion, manufacturers have nearly disposed of this code. The differing lead length has become merely fashion. And worse, what about the LEDs found in surplus, special-application LEDs that have both leads cut the same exact length? How can you possibly know the polarity then? Easy. All common LEDS have a similar internal structure. Look inside the dome. One lead will terminate in a small post. The other will terminate in a large dish of metal. The LED element rests above this large chunk of metal, and this dish represents the cathode, or negative side, of the LED. In the projects that depict LEDs, you'll see a dot on one side of the LED. This indicates the cathode, or large metal dish side, of the LED, regardless of lead length.

Note If the LED doesn't light, forget everything I just said and reverse the leads.

- **Circuit standards (in Appendix A):** Rather than draw the generic output scheme for each project (line out, trimmer, LED, speaker switch, and output jack), I've presented this common circuit in Appendix A. It's one of a dozen generic bends that can be applied to circuit-bending projects, including many in this book. Simply look at the output scheme in the appendix and consider the speaker shown to be the speaker in your current project. Plan your output controls (jack, speaker switch, and so on) and their placement with this general circuit in mind.

Note Check Appendix A for instrument output wiring and other common circuits.

Parts: Advice and a Warning

As covered early in this book, I highly advise buying parts in quantity from a surplus electronics dealer. Appendix C provides you with a few sources. Because many people will be buying parts from Radio Shack (after all, I've supplied you with RS part numbers), I must in good conscience sound a warning.

Though expensive, most Radio Shack LEDs, potentiometers, capacitors, and such will serve the purpose. But I'm not about to give any a reward for dependability. Radio Shack's tiniest "micro mini" toggle switches tend to fall apart or jam after a while, and their normally closed mini pushbutton switches are simply worthless. Avoid!

These pushbuttons are lacking a good crimp where the metal body meets the plastic back half. The result? The switch is guaranteed to jettison its bottom section within normal use, if not simply break in half as you attempt to mount it. Things get worse. There is no excuse for the poor machining of either the threaded shaft of this switch or the panel nut that is supposed to screw onto it, unless the low-quality soft metal used in each is to be the scapegoat. The result is a nut too loose for the job, and one that's difficult to thread onto the switch's shaft. The resulting jam will deform the soft metals, making the threading-on of the nut all but impossible. Again, avoid!

Perhaps Radio Shack will someday replace this design with a viable version. But as of this writing, and for years prior, Radio Shack has been peddling this certain inferior component. Trust me: Even if you don't have to pay shipping to and from Japan incurred by replacing a sloppy Radio Shack switch (the voice of experience speaking), you'll be frustrated enough with this switch to curse the day you bought it.

I'm stressing this point for one reason: You'll be needing a normally closed pushbutton switch for the power reset circuit in many projects. Radio Shack will be, even with this warning, the inevitable source for this switch for many benders. All I can do is try my best to help people circumvent all the expense, time wasted, and sure hassles this particular component brings.

My suggestion is to contact either Mendelson's or Pembleton Electronics (below) and buy an assortment of small components such as you'll need in these projects to see what you think. If you like what you get, order a decent quantity of all the common parts you'll need.

Describe what you want exactly—give dimensions and Radio Shack part numbers if they'll cross reference for you. Ask whether you're buying surplus or catalog items. Ask the price differences per piece and in quantity between catalog and surplus—ask where the price breaks kick in. Ask about the quality of each (often surplus is of better quality, consisting as it does of major corporation overruns). Ask about return policies if you're buying in quantity (you might find a defect the reseller honestly didn't know about). If you discover a defect in surplus equipment, report it to the seller immediately. You'll usually get a credit or your money back if the complaint is legit and you're friendly about it, even if your receipt says "All Sales Final."

You know what? Go ahead and buy the bad switch from Radio Shack. Seriously. Just one pack. The part number is 275-1548. But *do not* install one in an instrument. With the first switch, just try to break it in half. That was easy. With the second switch, just try to mount it in a hole and thread the nut on the switch. That was hard; notice the loose fit. Compare its construction and its hardware fit with that of a solid, well-built switch. I can think of no better lesson in identifying component quality than this! There's big edification here if you're willing to indulge.

The bender's last word on parts? Think small. Think miniature potentiometers (as long as they're panel-mount style and come with mounting hardware), small switches of all kinds (panel-mount again), small pilot lamp housings, and tiny versions of everything you'll ever want to use as a control. Why? Because limited space will become your constant companion. The two of you will need to make friends and get along really, really well.

How Are the Projects Arranged?

The eighteen projects are arranged in three general groups, and within each group you'll encounter an increasing range of difficulty as you proceed. The first group is devoted to Incantors and includes four projects. The next group is that of Aleatrons and consists of six projects. The last group encompasses eight projects in the area of who-knows-what, representing the kind of oddball circuits the bender is always facing. If you're interested in smoothing the learning curve, tackle this last group first, beginning with Project 11, the Mall Madness Machine.

Whose Is It, Anyway?

My philosophy is, if you build it, it's yours. But that's not where philosophy ends. I present the following projects for personal education and enjoyment, not for resale. These instruments are my own discovery, all original, and an outgrowth of nearly 40 years of study. Let them fascinate you; let them teach you. Let them serve as an example of what's possible within the field. Beyond that, let them be.

If you're interested in sales, do what I did: Discover your own original designs. Be an honest designer and don't knock off anyone else's circuits for business purposes! This book, along with your own experimentation, will surely lead you to many entirely original instruments. Take it from a seasoned designer: There is no other way to go.

Part Numbers

In the following section, I've listed all the part numbers for the electrical components needed to complete the projects in this book. Most are Radio Shack catalog numbers except when Radio Shack does not carry the product or the Radio Shack version is of too poor quality to use.

When contacting surplus outlets, know (as mentioned) that they generally carry two kinds of merchandise. One will be a catalog item (new) and the other will be whatever they currently have as true surplus in stock, sometimes used and removed from service and at other times never used but from manufacturer overruns from a while (sometimes a very long while) ago. Ask for descriptions.

Call or fax before you order and be sure that you know what you're buying. And as I said before, ask where the price breaks are and buy in quantity when you find a good thing.

I swear by surplus, but I'm still stuck here with 500 great-looking brushed-metal knobs that don't quite fit the ¼" shafts they were made for, their shaft holes milled out just a fraction too small. No wonder they ended up in surplus, eh? Well, my drill press solved that. Surplus *is* where it's at. Just be cautious.

Master Parts List and Sources

The projects in this book require very simple parts. They're commonly available and easy to understand. The only "foreign language" you'll see in this list describes the potentiometers (pots). And this language is limited to two letters only: K and M.

Potentiometers, as you learned earlier in this book, are variable resistors. The chief difference in pots is the resistance they can present to the electrons you have running through them. Electrical resistance is measured in ohms (Ω). Once into the thousands you'll find a K in the name (500K, or kilohm), and further, into the millions, you'll see an M in the title (10M, or megohms). That's all these symbols represent. Just be sure to observe these resistance values while building the projects and buying the pots you need.

Radio Shack:

- Mini Toggle Switch (SPDT), RS part # 275-613
- Mini Toggle Switch (SPST), RS part # 275-612
- Mini Pushbutton, momentary, (NO), RS part # 275-1547
- Sub–Mini Pushbutton, momentary, (NO), RS part # 275-1571
- Wire Wrap Wire, RS part #278-501
- Bus wire, RS part # 278-1341
- Red LED, RS part # 276-307
- Blue LED, RS part# 276-316
- White LED, RS part # 276- 320
- Photo Cell, RS part # 276-1657
- 5K pot, RS part # 271-1714
- 10K pot, RS part # 271-1715
- 50K pot, RS part # 271-1716
- 100K pot, RS part # 271-092
- 1M pot, RS part # 271-211

- DC Plug, RS part # 274-1573

- DC Jack, RS part # 274-1576

- Banana Jacks, RS part # 274-725b

- Banana Plugs, RS part # 274-730

Pembleton Electronics:

- Mini Pushbutton, momentary (NC), part # GC 35-3458

- Bending Probes, part # GC 12-1647

CE Distribution:

- Mini (16mm) 500K pot for Incantors, part # R-VAM-500KA

Your local hardware store:

- Body-contact (tapped 8/32 brass ball, Westinghouse part # 70660 or equivalent). You'll find it in the lamp or home lighting section.

Note
I have assumed that you have at hand rosin core solder, wire-wrap wire, bus wire, eyelets for your body-contacts, heat-shrink tubing, and the proper workshop tools outlined previously in this book. These items, which are common to all projects, do not appear in the project parts lists.

Contacts:

- Radio Shack can be contacted through its walk-in retail outlets or via www.radio shack.com.

- Pembleton Electronics can be faxed at 260-484-0163 or called at 260-484-1812.

- CE Distributors can be called at 1-800-840-0330; website: www.cedist.com, email: info@cedist.com

Several projects call for potentiometers with a value above 1M. These can be difficult to find. If there's an electronic surplus outlet in your area, pay it a visit. If not, search online, and give the following retail outlets a try:

- http://www.webervst.com/pots.html

- https://weberspeakerscom.secure.powweb.com/store/potsord.htm

- http://www.meci.com/

- www.allparts.com

- electronicsurplus.com

- debcoelectronics.com

- www.allelectronics.com
- http://www.kenselectronics.com

Pembleton Electronics (listed previously) might also have large value, panel-mount pots in stock. Give them a call.

Note

Be sure to check out Appendix B for more bending resources.

There. Now we're covered. Here comes the fun stuff. . . .

Project 1: Original Pushbutton Speak & Spell Incantor

Sir Richard Paget, an insightful scientist of the early 1900s, somehow managed to get hold of a complete human vocal tract removed from a corpse destined, as it were, to speak again. Paget's work with this human tissue of tongue, throat, and larynx followed a long and, at times, macabre history of investigation into, first, understanding the mechanics of speech, and second, synthesizing it.

Speech, after all, is an ultimate indicator of thought (even if Polly doesn't *really* want a cracker). What is the talking dummy without the ventriloquist? The intrinsic force of speech itself as ultimate communicator forces its cerebral impact and intellectual power, even if the speaker is unlikely or inanimate. Or dead.

While an assistant directed compressed air through the vocal tract, Sir Richard manipulated the cold flesh, creating sounds that not only illustrated vocalization processes but also ended up inspiring a utilitarian invention. The vocal tract, now molded in plasticine and combined with resonators, was reconfigured as a motor buggy horn enabling the deceased to shout "Away! Away!" at the living. In this case the living were often horses, already frightened by the horseless carriage and hardly in need of further, dare I say, spooking.

Attempts to harness or synthesize the power of human speech reach way into the past, usually with amusing results. Greek deities such as the Speaking Head of Orpheus contained hidden tubes through which divine (or at least priestly) communications could pass. In Cervantes' *Don Quixote* the conversing bust of a Roman emperor was likewise endowed. But the key to the final unlocking of synthetic speech would have to wait until the advent of electronics, telecommunication, and, in fact, the birth of Homer Dudley.

The Bell Telephone Laboratories introduced Dudley's invention in 1937. Known as the VODER, this wrap-around console, replete with foot pedals and piano-like keys, was capable of producing somewhat convincing human speech. You and I would have to wait decades, however, to play around with electronic synthetic speech technology. We'd have to wait until 1978, when the Texas Instruments Company released what would become the most circuit-bent instrument in the world, the Speak & Spell.

The first Speak & Spell had pushbuttons and a three-wire speaker (see Figure 14-1). I bent it immediately, naming it the Feletcress Solution Device. The Speak & Read and Speak & Math would follow, with the entire series proving to be wonderfully bendable.

I introduced my findings in 1993 in *Experimental Musical Instruments* magazine, naming the bent instrument the "Incantor" (as in incantation). You'll be building all four Incantors here, unlocking one of the most interesting alien music engines around.

Bending all versions is straightforward. Note, however, that the battery compartment has a pair of spring contacts at one end, coaxing you to insert the four batteries all pointing in the same direction. Fine, if you want a foot warmer. But if you don't want the batteries to fry, position them the right way, contrary to what the battery contacts seem to indicate: two facing one way and two the other, just as the easy-to-miss diagram molded into the bottom of the plastic compartment advises.

In some models, you'll discover a speaker held in position only by the back half of the case. You'll need to keep this speaker in place while you work. A few dots of hot-melt glue will do the trick, as will a section of masking tape.

Keep in mind that there are many variants of the circuits contained within the Speak & Spell product series. If your model does not exactly match the circuit shown in Figure 14-2, don't worry. Bending points are often the same. Just follow the diagrams and wire things as shown. If the diagram indicates that a bending wire attaches to point 3 in a line of 10 similar points, solder your wire to point 3 even if the actual printed circuit traces around the 10 points look different. If a connection doesn't have the result intended, poke around in the same area on the circuit. A similar bend is often near.

In your rummaging around, you're likely to find, in time, both the original pushbutton Speak & Spell as well as the later version with a flat, membrane keypad. Let's look at bending the original version first.

The original pushbutton Speak & Spell, although the classic, is the least hi-fi of the series. If you like the coarseness of early speech synthesis, this unit is for you. Its fidelity was still groovy enough to inspire the later variants (Speak & Read and Speak & Math), both sounding much clearer.

You'll do a double-take when you pop the original Speak & Spell open and look at the speaker. Three wires? In this first Speak & Spell, the vocalizations were split between two channels. These two signals were mixed at the speaker, the extra signal using the extra wire. That's pretty weird. You'll have fun with it.

Body-Contact   Pitch Dial   Looping Switches

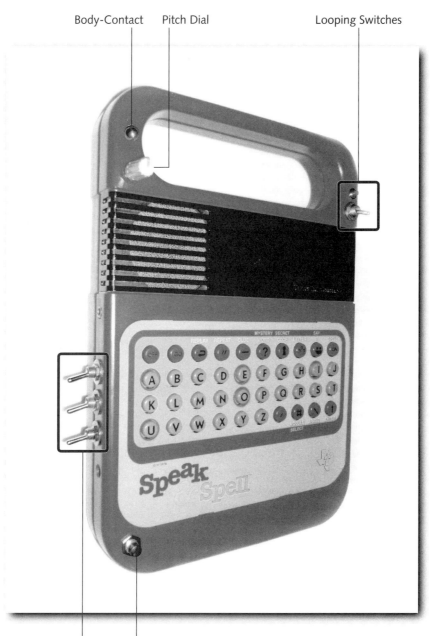

Bending Switches   Reset

FIGURE 14-1: Original Speak & Spell in standard Incantor configuration

But before you do anything, get familiar with the unit's operation by playing all the games. When you know your way around the Speak & Spell's various operations and have your bending skills practiced and ready, it's time to dive in.

Bending a "Speak" requires a little work, but you'll be rewarded with a very intriguing alien instrument when you're done. Just be thankful that Sir Richard paved our way and that today we can experiment with synthetic speech without robbing a single grave.

Parts

Before you begin this project, be sure to have all these parts at hand.

- 1 pushbutton-style Speak & Spell
- 4 miniature SPDT (or SPST) toggle switches
- 1 miniature N.C. (normally closed) pushbutton switch
- 1 sub-miniature N.O. (normally open) pushbutton switch
- 1 miniature 500K potentiometer
- 1 body-contact

Open It Up!

Grab your tools and let's see what's inside!

1. Remove battery door and batteries.

2. Remove the two screws at the bottom of the back (these will be either Phillips or star [Torx] head). Set screws aside on your hardware magnet or some other safe place.

3. Spread case halves apart by prying with your fingers inside the battery compartment.

4. With case spread a little bit, insert a screwdriver in the lower of the four rectangular holes on the case back and gently push the plastic retaining clips to the side. The case will pop open a bit as each clip is released.

5. Do the same with the top two clips and set the back of the case in a safe place.

Circuit at First Glance

You're looking at the back of the circuit board. That's just fine—all you need to get to is right there. You'll see the power and headphone jacks on the side and the battery compartment at the bottom. The long parallel rows of soldered contacts are mostly IC pins. These contacts are what you'll be soldering to.

Is there an expansion cartridge plugged into the circuit edge? Usually accessed through the battery compartment, the cartridge port may or may not be occupied. If it is, remove the cartridge and set it aside. After bending, your new Incantor will accept the expansion cartridge once again, and even have something to say about it. Many different cartridges were produced for the Speak series. With luck, you'll collect a set to use with your instruments.

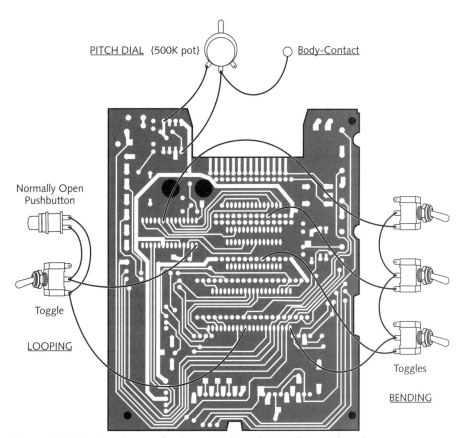

FIGURE **14-2: Bending diagram for the original Speak & Spell circuit board**

Step-by-Step Bending

You've gathered all the needed parts and you've opened things up. You've studied the diagram. Now, step by step, you'll complete the transformation. Here's how.

Choosing a Cool Control Layout

Inside the Speak & Spell you'll find room enough to mount controls without too much difficulty. Pictured is the standard control configuration I go with most of the time (refer to Figure 14-1). If you want to vary this, feel free. You'll find more room above the keyboard if you remove the circuit board to get there.

Keeping the two looping switches close to each other is a good idea. Same goes for the three bending switches. This makes things easier to wire.

The reset switch should be mounted in a spot not easily hit by accident. I have it pictured on the top surface of the instrument, away from other controls and in a spot where it's very simple to mount. If you think you'll hit it too easily there, try mounting it on the instrument's side, perhaps to the bottom left of the circuit board. Side mounting is not as easy as top surface mounting due often to switch insertion angles and general clearances. But it's possible in the right spot. As always, be sure of all clearances before drilling a hole for any component.

Case Considerations

The plastic used in all "Speaks" is sturdy as well as elastic enough to be drilled without cracking. There are usually no ribs to remove, though in this model there might be.

If this (or any) Speak has a plastic cylinder on the backside of the case, opposite the speaker, take a close look at how you run your new wires near the speaker. If you think the plastic cylinder will interfere with your wires, either run the wires differently or clip away enough of the plastic to keep it from touching the wires (if you intend to solder wires to the speaker for any reason, you'll need to consider this issue).

In this first model, Texas Instruments did use support pillars inside the case to keep the keyboard pressed against the front of the housing. One or two of these pillars might hit the switches if mounted along the side of the case as shown. Using a strong wire clipper, just clip the troublesome support pillar away from the back edge of the case.

Marking the Board

Using your permanent ink, ultra-fine-tipped Sharpie pen, mark the board to indicate the soldering points shown in the bending diagram in Figure 14-2. Put a tiny dot next to each point to be soldered to.

Drilling Holes

After you've checked clearances and decided on a control layout, mark hole positions with your pencil.

All holes start with your pilot bit of ⅛" or so in diameter. Most critical here is the spacing of the three bending switches. If you'll be placing them as in the illustration, you must be careful to keep them equidistant, and between the headphone and power jacks. Mark the position of the middle switch first and work off of that. Be sure that the switches are mounted just high enough to clear the back of the pushbutton keyboard. If the collars of the switches interfere with the closing of the case later, using a small wire clipper just snip a little of the back side's rim away opposite each collar.

When your pilot holes are drilled, your next step is to bring them up to the correct diameter with your hand bore. As you use the bore keep checking the hole size against the component you're mounting. The accessible hole edges should be finished with the de-burrer. Any extra plastic at the edges of hard-to-get-to holes can be scraped away with a small knife or screwdriver blade.

Painting

After all holes are correctly sized, you're ready to paint. Remember, if the instrument is to be painted you'll need to make the holes a little larger than otherwise (see Chapter 12).

You'll probably want to use ½" masking tape to mask the keypad, the jacks on the edge of the circuit board, the speaker, and the fluorescent display. While you're masking the keypad, concentrate on the edge of the tape running along the outside edge of the pad. Turning the corners will be easy this way—the excess inner tape will fold upon itself while you turn corners. Whether at a curve or on a straight line, firmly press masking tape down bit by bit as you go.

Fill in the central area with either strips of masking tape or, better yet, a piece of newspaper folded to size and taped down to cover the rest of the keyboard.

To avoid the paint's cracking, you should remove the tape before the paint is totally dry. Exactly when to do this depends on the individual paint job. It's usually okay to remove the tape within a few hours of spraying. Be careful here. You'll have to touch only unpainted areas of the instrument because the still-tacky paint will mar easily.

If you've painted your instrument, be sure that all paint, titles, and gloss coats are dried completely before proceeding. Review Chapter 12 for painting and drying tips.

Control Mounting

The mounting of controls is straightforward as long as you've obtained a miniature 500K potentiometer (the usual 1"-diameter pot is too wide to fit in the space available; see parts list for the correct 16mm-diameter pot). Mount all components.

Soldering

And now for the fun part! If you've followed the guidelines in the soldering section (Chapter 8), and tried all the techniques on a practice circuit board to the point that you're comfortable, then you're all set. If not, *do* return there and sharpen those skills.

Using your bus wire, connect together the middle lugs of the three bending switches (refer to Figure 14-2). Be sure that the wire touches only the soldering lugs intended and nothing else.

Using your insulated wire-wrap wire, finish this three-switch matrix by soldering the indicated switch lugs to the circuit points shown. You'll be soldering four wires here: one to the three lugs you just connected, and one to each of the three switch's lower lugs.

Next, wire the two switches of the looping matrix, exactly as shown. Wire the pitch dial as shown. Wire the body-contact as shown.

Wire the reset switch by snipping one of the wires coming from the battery compartment and soldering your normally closed pushbutton switch in the middle (see Appendix A for reset button wiring).

Note Consider mounting a miniature stereo headphone output on the instrument. Solder extension wires coming from the three speaker wires to this jack, starting with the common (-) side of the jack going to the negative wire attached to the speaker (only one of the speaker's wires goes to a negative terminal; see markings at speaker terminals). Wire the remaining two positive (+) wires to the two open positive lugs on the stereo jack. You now have a stereo Incantor! You'll get different results by varying this three-wire scheme. It won't hurt to switch wires around on the stereo jack and listen to what happens.

Testing

You now have an octopus-like thing in front of you, wire tentacles waving and Cyclops speaker eye staring you down. Let's see what it does.

1. Turn the pitch dial all the way up and turn all your added toggle switches off.

2. Install fresh batteries.

3. Holding the batteries in, press the Speak's ON button. If all has gone well, you should hear the usual start-up sequence. If not:

   a. Be sure that you've used the right switches in the right places (your N.O. and N.C. pushbuttons).

   b. Be sure that you've wired things correctly (check that the correct lugs on pot and toggle switches are connected as in the illustration),

   c. Be sure that the batteries are in correctly (two one way and two the other) and are making good contact.

Troubleshooting Tips

Don't be alarmed if a bent project doesn't work when it's first turned on. This happens a lot. Usually the problem is sitting right there in front of you, with just a little poking around needed to find it.

1. Hold the reset switch in as you start up. If this works, you've used a normally open instead of a normally closed pushbutton. Install the correct switch.

2. Turn the pitch dial most of the way down while you start up. If this works, you've wired the pot in reverse. Remove the wire from the outside lug and solder it to the other outside lug.

3. Do both of the preceding at the same time while you start up. If this works, install the correct pushbutton and reverse the wiring on the pot, as noted.

   If you're still having no luck, recheck all wiring for overdone solder joints or incorrect connections. When you hear the start-up sequence and the instrument is operating normally, you can proceed with testing.

4. Test the reset switch. While the Speak is speaking, press the reset switch to confirm power supply interruption.

5. While the Speak is in the middle of a routine (try all of its game functions), press and hold the pushbutton looping switch. If you're in the right routine, the sounds should begin to loop. Press and hold again for another loop, restarting a game routine if necessary should the Speak fall silent (or quickly hit the reset switch if the Speak crashes).

6. When you find a good loop, flip the looping toggle switch to lock the loop in place.

7. Try slowly turning the pitch dial down. Be careful when you see the display begin to shudder. This means that you're at the bottom end of the pitch range and turning the dial further will result in a crash. Touching the body-contact connected to the pitch dial should produce vibrato and pitch bending.

Note See Chapter 9, Figure 9-23, as well as Appendix A, for instructions on adding a trimmer pot to the main pot to keep the main pot from crashing the circuit. You'll lose a little of the ultra-slow clocking range if you install a trim pot, but you'll never crash the bottom end of a loop again.

8. Last to test are the three bending switches. With pitch turned all the way up and the Speak doing one of its normal routines, try flipping one of the three bending switches. If nothing happens, try another of the Speak's routines. If it's on the right routine, the instrument should begin to produce streams of wild sound elements, some lasting a little while, some going on and on. These switches should all produce similar yet different disruptions, singly or in combination. Experiment with the Speak's data input keys (ON, Enter, and so on) to see whether you can start a new sequence while a bending switch is turned on.

Tip Remember to either immediately hit the reset switch or the power switch or to quickly remove a battery upon the crashing of any bent instrument!

Reassembly

Simple. Just be sure that all wires are out of the way so that they're not pinched between the circuit board and the back of the housing. Align the back of the housing with the plastic clips that you pried apart initially. Press sides together firmly and they'll snap into place.

As noted previously, should the plastic support pillar opposite your new switches interfere with closing the case, carefully snip the pillar away with a sturdy wire clipper. And if the edge of the back of the case hits the collars of the three bending switches, snip a little of the edge away opposite each collar.

Replace the two screws at the bottom of the case and you're done. Recheck all circuit functions.

Musicality

Looping is immediately musical and can be accompanied with many other instruments. Incantor looping is fascinating not only in obvious individual loop output but also in the regard that the looping function is bottomless. That is, like a bottomless well, there is no end in sight to the loops achievable. You will always be able to find new musical cycles.

Slowing down sounds via the pitch dial brings musical tones out of even ordinary speech. The streams of abstract noises via the bending switches, at least to my ears, present as musical entities immediately. And the body-contact will allow the player a controlled vibrato, the classic animation of so many musical voices.

Unique to the Speak & Spell models is the response of the keyboard with the bending switches thrown. You'll find that some of the letter keys create cool sounds when pressed. Press the same key again and you'll get a different sound, the second in the pair that a key is now able to produce. Not all keys respond in this way, and the later models in the series (Speak & Read, Speak & Math) do not exhibit this response (though some models will repeat sounds when keys are pressed—experiment!).

Incantors are great for sampling and running through outboard effects (use the headphone output). I like the effect of blending two (or more) live Incantors into each other, and the unpredictable combinations that result. Some can be very intriguing.

Project 2: Common Speak & Spell Incantor

Membrane keypads are much simpler, less expensive, easier to make, and less breakable than the pushbutton keyboard found on the first Speak & Spell. Accordingly, Texas Instruments abandoned the pushbutton keyboard in favor of the membrane pad soon after the original Speak & Spell was first marketed.

Most of the Speak & Spells you'll find are of this second style (see Figure 15-1). They also sound better, using a revamped voice synthesizer and finally a traditional two-wire speaker.

But prior to doing anything, as before, get familiar with the unit's operation by playing all the games.

Parts

Before you begin this project, be sure to have all these parts at hand.

- 1 membrane keypad Speak & Spell
- 4 miniature SPDT (or SPST) toggle switches
- 1 miniature N.C. (normally closed) pushbutton switch
- 1 sub-miniature N.O. (normally open) pushbutton switch
- 1 miniature 500K potentiometer (see master parts list in Chapter 13)
- 1 body-contact

Pitch Dial   Body-Contact   Looping Switches

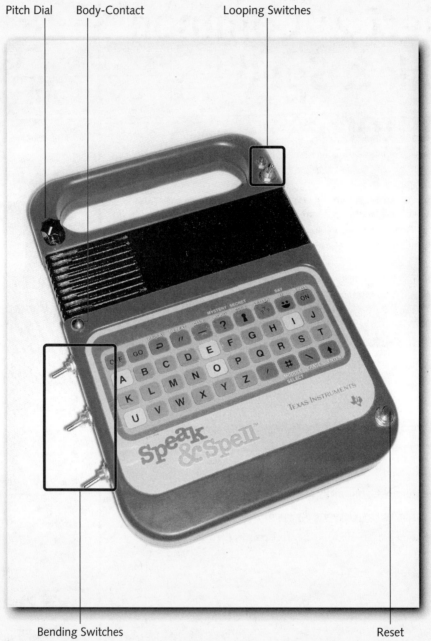

Bending Switches   Reset

FIGURE 15-1: Common membrane-keypad Speak & Spell in standard Incantor configuration

Open It Up!

Follow the instructions for opening the original Speak & Spell, covered previously in Project 1 (Chapter 14). The procedure is exactly the same.

Circuit at First Glance

For us, the main difference between the revised Speak & Spell and the original is the circuit board: It's considerably smaller (see bending diagram, Figure 15-2). As mentioned, you're also now looking at a traditional two-wire speaker, the kind you'll find in all the other projects in this book.

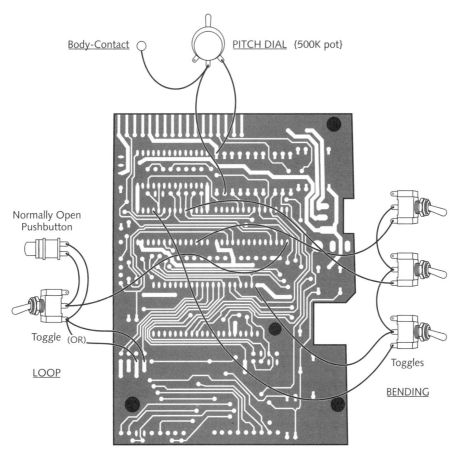

FIGURE 15-2: Bending diagram for the common, membrane-keypad Speak & Spell circuit board. Note alternate soldering point for one of the two looping connections. If the main point doesn't work, try the alternate.

Is there an expansion cartridge plugged into the circuit edge? Usually accessed through the battery compartment, the cartridge port may or may not be occupied. If it is, remove the cartridge and set it aside. After bending, your new Incantor will accept the expansion cartridge once again, and even have something to say about it. Many different cartridges were produced for the Speak series. With luck, you'll collect a set to use with your instruments.

Step-by-Step Bending

You've gathered all the needed parts and you've opened things up. You've studied the diagram. Now, step by step, you'll complete the transformation. Here's how.

Choosing a Cool Control Layout

Because body design became the same throughout the Speak & Spell series after the original pushbutton model was replaced, you can mount components the same way each time throughout the rest of the series, whether working with a Speak & Spell, Read, or Math. And this time you'll have more choices.

Case Considerations

The main difference now is in the mounting of the set of three bending switches. Although you certainly can mount them on the side of the instrument as I described with the pushbutton Speak & Spell, you now also have the option of mounting them next to the speaker (the larger printed circuit on the prior model disallowed this). In this position they'll be a little bit handier (refer to Figure 16-1, next chapter).

If you choose to opt for this alternative switch-mounting scheme, removal of the plastic lens in front of the speaker is a good idea. Drilling the mounting holes for the switches will be easier with the lens removed. And if your toggle switches have short, threaded collars, you'll now have more collar for the switch nuts to screw down onto.

To remove the lens, just pry it away from the case where it ends its wrap around the side. Use a small screwdriver to do this. The lens will pop off without much trouble.

On some Speaks, the actual speaker is covered with a black grill cloth. That's good. On other versions the speaker cone is totally exposed. Gluing a grill cloth over the exposed speaker is a nice idea (see Chapter 11, Figure 11-22).

As indicated in the discussion of the original model (Project 1; see Chapter 14), take a look at the area opposite the speaker on the back half of the case. If you see a cylinder of plastic meant to back the speaker when the case is closed, be sure that no wires are pinched by this cylinder when you close the finished instrument.

Marking the Board

Using your permanent ink, ultra-fine-tipped Sharpie pen, once again mark the board to indicate the soldering points shown in Figure 15-2. Put a tiny dot next to each point to be soldered to.

Drilling Holes

All holes start with your pilot bit of ⅛" or so in diameter. Most critical here is the spacing of the three bending switches. If you'll be placing them as in Figure 16-1 in the next chapter, you must be careful to keep the switches from hitting the speaker or each other. Mounting these three switches along the edge of the housing is, again, another possibility (on the speaker side of the instrument), just as you did with the pushbutton Speak & Spell covered in Project 1. Bring the holes up to size with your hand bore and finish their edges with the de-burrer.

Painting

Once again, all the same considerations apply as with the original Speak & Spell covered previously, in Project 1.

Control mounting

As before, mount all components prior to soldering. Be sure that you're using a normally closed pushbutton switch for the reset switch.

Soldering

Using your bus wire, connect together the middle lugs of the three bending switches (refer to Figure 15-2). Be sure that the wire touches only the soldering lugs intended and nothing else.

Using your insulated wire-wrap wire, finish this three-switch matrix by connecting the indicated switch lugs to the circuit points shown. You'll be soldering four wires here: one to the three lugs you just connected together, and one to each of the three switch's lower lugs.

Next, wire the two switches of the looping matrix, exactly as shown. Wire the pitch dial as shown. Wire the body-contact as shown.

Wire the reset switch by snipping one of the wires coming from the battery compartment and soldering your normally closed pushbutton switch in the middle (see Appendix A for reset button wiring).

Note

You might look for envelope LED bends as well. Several usually exist. Know, however, that some LED bends will keep the circuit from operating correctly depending on the version of the Speak & Spell you're working on and where you wired the bend. If the circuit seems to stall or not start during the testing section that follows, remove the LED from the circuit and begin the test routine again. If the LED burns too brightly, add a 100 ohm resistor (brown-black-brown color code) to its anode (the *usually* longer lead and always the one terminating inside the LED as the smaller metal mass). Always experiment with various resistors when adjusting LED brightness.

Testing

So far, so good. Let's put this thing through its paces.

1. Turn the pitch dial all the way up and turn all your added toggle switches off.

2. Install fresh batteries.

3. Holding the batteries in place, press the Speak's ON button. If all has gone well, you should hear the usual startup sequence. If not:

    a. Be sure that you've used the right switches in the right places (your N.O. and N.C. pushbuttons).

    b. Be sure that you've wired things correctly (the correct lugs on pot and toggle switches are connected as in the illustration).

    c. And be sure that the batteries are in correctly (two one way and two the other) and are making good contact.

Troubleshooting Tips

Don't be alarmed if a bent project doesn't work when it's first turned on. This happens a lot. Usually the problem is sitting right there in front of you, with just a little poking around needed to find it.

1. Hold the reset switch in as you start up. If this works, you've used a normally open instead of normally closed pushbutton. Install the correct switch.

2. Turn the pitch dial all the way down while you start up. If this works, you've wired the pot in reverse. Remove the wire from the outside lug and solder it to the other outside lug.

3. Do both of the above at the same time while you start up (dial down; switch held in). If this works, install the correct pushbutton and reverse the wiring on the pot, as noted.

   If you're still having no luck, recheck all wiring for overdone solder joints or incorrect connections. When you hear the startup sequence and the instrument is operating normally, you can proceed with testing.

4. Test the reset switch. While the Speak is speaking, press the reset switch to confirm power supply interruption.

5. While the Speak & Spell is in the middle of a routine (try all of its game functions), press and hold the pushbutton looping switch. If it's in the right routine, the sounds should begin to loop. Press and hold again for another loop, restarting a game routine if necessary should the Speak fall silent (or quickly hit the reset switch if the Speak crashes).

6. When you find a good loop, flip the looping toggle switch to lock the loop in place.

7. Try slowly turning the pitch dial down. Be careful when you see the display begin to shudder. This means that you're at the bottom end of the pitch range and turning the dial further will result in a crash.

Note

See Chapter 9, Figure 9-23, as well as Appendix A, for instructions on adding a trimmer pot to the main pot to keep the main pot from crashing the circuit. You'll lose a little of the ultra-slow clocking range if you install a trim pot, but you'll never crash the bottom end of a loop again.

8. Last to test are the three bending switches. With pitch turned all the way up and the Speak doing one of its normal routines, try flipping one of the three bending switches. If nothing happens, try another of the Speak's routines. If it's on the right routine, the instrument should begin to produce streams of wild sound elements, some lasting a little while, some going on and on. These switches should all produce similar disruptions, singly or in combination.

Tip

Remember to hit the reset switch, turn off the power switch, or quickly remove a battery upon the crashing of any bent instrument!

Reassembly

Same as before with the first model I discussed in Project 1 (Chapter 14). Just snap the back on and be sure to keep all wires from being pinched as you close the case. See Project 1 for details.

Musicality

Because all Incantors output the same basic signals, musicality is generally the same. Read the previous description of the original Speak & Spell's output for insights in Project 1.

Project 3: Speak & Read Incantor

Just as before, this next "Speak" in the series further improved speech technology and circuit design (see Figure 16-1). Not only does it sound better, but the ROM containing speech holds the most words in the entire series, including the next model to follow, the Speak & Math.

Parts

Before you begin this project, be sure to have all these parts at hand.

- 1 Speak & Read
- 4 miniature SPDT (or SPST) toggle switches
- 1 miniature N.C. (normally closed) pushbutton switch
- 1 sub-miniature N.O. (normally open) pushbutton switch
- 1 miniature 500K potentiometer (see master parts list in Chapter 13)
- 1 body-contact

Open It Up!

You're an old hand at this by now. Get familiar with how the unit is supposed to work (play all the games) and then follow the instructions for opening the original Speak & Spell, as covered in Project 1 (Chapter 14).

Circuit at First Glance

Pretty similar to the Speak & Spell just discussed. At times, the little circuit off to the side of the main board will differ from version to version. Don't worry—this won't affect your work. See the bending diagram in Figure 16-2 to locate the most useful Speak & Read bends.

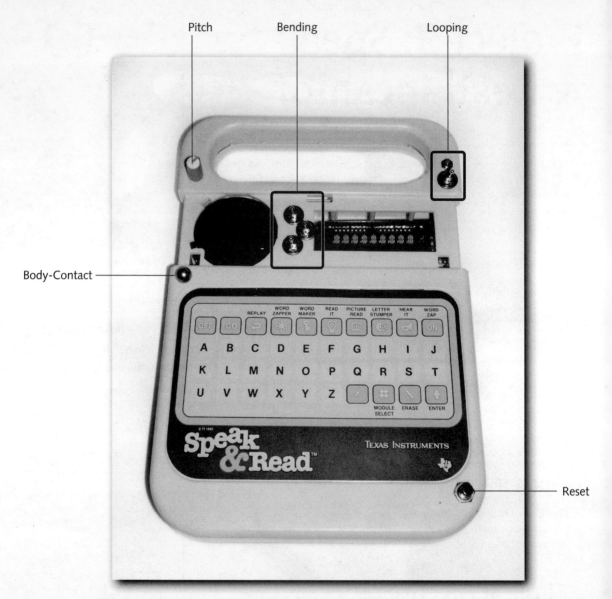

FIGURE 16-1: Speak & Read in standard Incantor configuration

Is there an expansion cartridge plugged into the circuit edge? This is usually accessed through the battery compartment; the cartridge port may or may not be occupied. If it is, remove the cartridge and set it aside. After bending, your new Incantor will accept the expansion cartridge once again, and even have something to say about it. Many different cartridges were produced for the Speak series. With luck, you'll collect a set to use with your instruments.

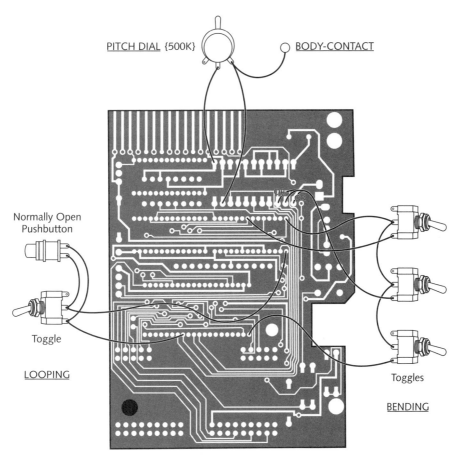

PITCH DIAL {500K}   BODY-CONTACT

Normally Open
Pushbutton

Toggle

LOOPING

Toggles

BENDING

FIGURE 16-2: Bending diagram for the Speak & Read circuit board. Note alternate soldering points for middle bending switch. Using the wrong point will speed up the circuit or otherwise interfere with normal operation.

Step-by-Step Bending

You've gathered all the needed parts and you've opened things up. You've studied the diagram. Now, step by step, you'll complete the transformation. Here's how.

Choosing a Cool Control Layout

Because body design is the same as that of the membrane keypad Speak & Spell just covered (see Project 2, Chapter 15), you can mount components the same way if you like. If you want your Incantors to look or operate differently, now's the time to consider various control arrays.

As before, grouping the two looping switches together makes sense, as does grouping the three bending switches together (refer to Figure 16-1).

Case Considerations

Check out the comments given previously for the membrane keypad Speak & Spell (Project 2).

Marking the Board

Grab that tiny-tipped Sharpie marker and mark the soldering points, as shown in the bending diagram (refer to Figure 16-2).

Drilling Holes

Same as before. All holes start with your pilot bit of ⅛" or so in diameter. Most critical here is the spacing of the three bending switches. If you'll be placing them as in the main layout illustration (refer to Figure 16-1) you must be careful to keep the switches from hitting the speaker or each other. Mounting these three switches along the edge of the housing is, again, another possibility (on the speaker side of the instrument), just as I described with the pushbutton Speak & Spell covered in Project 1 (Chapter 14). Bring all holes up to size with your hand bore and finish with the de-burrer.

Painting

Once again, all the same considerations apply as with the original Speak & Spell covered previously, in Project 1.

Control Mounting

As before, mount all components prior to soldering. All controls should be easy to mount if located as shown in the main layout illustration (refer to Figure 16-1).

Soldering

Using your bus wire, connect the middle lugs of the three bending switches (refer to Figure 16-2). Be sure that the wire touches only the soldering lugs intended and nothing else.

Using your insulated wire-wrap wire, finish this three-switch matrix by soldering the indicated switch lugs to the circuit points shown. You'll be soldering four wires here: one to the three lugs you just connected, and one to each of the three switch's lower lugs.

Next, wire the two switches of the looping matrix, exactly as shown. Wire the pitch dial as shown. Wire the body-contact as shown.

Wire the reset switch by snipping one of the wires coming from the battery compartment and soldering your normally closed push-button switch in the middle (see Appendix A for reset button wiring).

Note You might look for envelope LED bends as well. Several usually exist. Know, however, that some LED bends will keep the circuit from operating correctly depending on the version of the Speak & Read you're working on and where you wired the bend. If the circuit seems to stall or not start during the testing section that follows, remove the LED from the circuit and begin the test routine again. If the LED burns too brightly, add a 100 ohm resistor (brown-black-brown color code) to its anode (the *usually* longer lead and always the one terminating inside the LED as the smaller metal mass). Always experiment with various resistors when adjusting LED brightness.

Testing

So far, so good. Let's put this thing through its paces.

1. Turn the pitch dial all the way up and turn all your added toggle switches off.

2. Install fresh batteries.

3. Holding the batteries in, press the Speak's ON button. If all has gone well, you should hear the usual start-up sequence. If not:

    a. Be sure that you've used the right switches in the right places (your N.O. and N.C. pushbuttons).

    b. Be sure that you've wired things correctly (the correct lugs on pot and toggle switches are connected as in the illustration).

    c. Be sure that the batteries are in correctly (two one way and two the other) and are making good contact.

Troubleshooting Tips

Don't be alarmed if a bent project doesn't work when it's first turned on. This happens a lot. Usually the problem is sitting right there in front of you, with just a little poking around needed to find it.

1. Hold the reset switch in as you start up. If this works, you've used a normally open instead of normally closed pushbutton. Install the correct switch.

2. Turn the pitch dial all the way down while you start up. If this works, you've wired the pot in reverse. Remove the wire from the outside lug and solder it to the other outside lug.

3. Do both of the preceding at the same time while you start up. If this works, install the correct pushbutton and reverse the wiring on the pot, as noted.

If you're still having no luck, recheck all wiring for overdone solder joints or incorrect connections. When you hear the startup sequence and the instrument is operating normally you can proceed with testing.

4. Test the reset switch. While the Speak is speaking, press the reset switch to confirm power supply interruption.

5. While the Speak & Read is in the middle of a routine (try all of its game functions), press and hold the pushbutton looping switch. If it's in the right routine, the sounds should begin to loop. Press and hold again for another loop, restarting a game routine if necessary should the Speak fall silent (or quickly hit the reset switch if the Speak crashes).

6. When you find a good loop, flip the looping toggle switch to lock the loop in place.

7. Try slowly turning the pitch dial down. Be careful when you see the display begin to shudder. This means that you're at the bottom end of the pitch range and turning the dial further will result in a crash.

 Note See Chapter 9, Figure 9-23, as well as Appendix A, for instructions on adding a trimmer pot to the main pot to keep the main pot from crashing the circuit. You'll lose a little of the ultra-slow clocking range if you install a trim pot, but you'll never crash the bottom end of a loop again.

8. Last to test are the three bending switches. With pitch turned all the way up and the Speak doing one of its normal routines, try flipping one of the three bending switches. If nothing happens, try another of the Speak's routines. If you're on the right routine, the instrument should begin to produce streams of wild sound elements, some lasting a little while, some going on and on. These switches should all produce similar disruptions, singly or in combination.

 Tip Remember to either hit the reset switch, turn off the power switch, or quickly remove a battery upon the crashing of any bent instrument! A delay in resetting a crashed circuit, depending upon the cause of crash, can result in burnout.

Reassembly

Same as before with the first model discussed in Project 1 (Chapter 14). Just be sure, as usual, to keep all wires from being pinched as you close the case.

Musicality

Because all Incantors output the same basic signals, musicality is generally the same. Read the previous description of the original Speak & Spell's output for insights, in Project 1. I especially like the bent Speak & Read's ability to string words together into amusing sentences. Don't be surprised if your model says "Let's smell the scissors s'more."

Project 4: Speak & Math Incantor

I f you're looking for the Incantor with the "hi-fi" voice, this is it.

Clearer sounding than any model before, the Speak & Math is also the most difficult of the later models to find (see Figure 17-1).

Parts

Before you begin this project, be sure to have all these parts at hand.

- 1 Speak & Math
- 4 miniature SPDT (or SPST) toggle switches
- 1 miniature N.C. (normally closed) pushbutton switch
- 1 sub-miniature N.O. (normally open) pushbutton switch
- 1 miniature 500K potentiometer (see master parts list in Chapter 13)
- 1 body-contact

Body-Contact   Pitch Dial   Bending Switches   Looping Switches

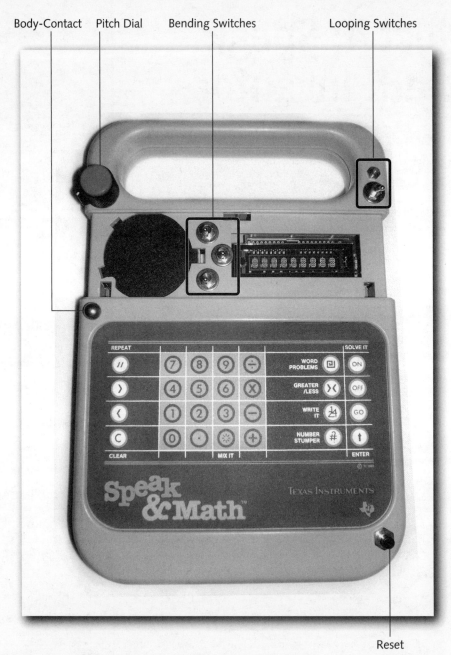

Reset

FIGURE 17-1: Speak & Math in standard Incantor configuration

Open It Up!

Get familiar with the unit's functions (play all the games). Then follow the instructions for opening the original Speak & Spell, covered previously in Project 1 (Chapter 14). I'll bet you guessed all that.

Circuit at First Glance

No surprises. Texas Instruments stuck to its internal layout quite rigorously, from the membrane-style Speak & Spell onward. Bending points are shown in Figure 17-2.

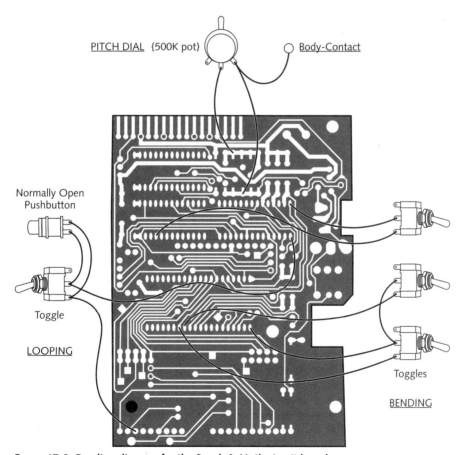

FIGURE 17-2: Bending diagram for the Speak & Math circuit board

Is there an expansion cartridge plugged into the circuit edge? Usually accessed through the battery compartment, the cartridge port may or may not be occupied. If it is, remove the cartridge and set it aside. After bending, your new Incantor will accept the expansion cartridge once again, and even have something to say about it. Many different cartridges were produced for the Speak series. With luck, you'll collect a set to use with your instruments.

Step-by-Step Bending

You've gathered all the needed parts and you've opened things up. You've studied the diagram. Now, step by step, you'll complete the transformation. Here's how.

Choosing a Cool Control Layout

Up to you! The usual Incantor layout (refer to Figure 17-1) will work fine for a Speak & Math. Vary as you like, keeping in mind that grouping the two looping and three bending switches together simplifies Incantor wiring.

Case Considerations

As you expected, check out the comments in this section for Project 2 (Chapter 15). The Speak & Math is no different.

Marking the Board

Using your tiny-tipped Sharpie, marker, mark the soldering points as shown in the diagram (refer to Figure 17-2).

Drilling Holes

Same as before. All holes start with your pilot bit of $1/8$" or so in diameter. Most critical here is the spacing of the three bending switches. If you'll be placing them as in the main layout illustration (refer to Figure 17-1), you must be careful to keep the switches from hitting the speaker or each other. Mounting these three switches along the edge of the housing is, again, another possibility (on the speaker side of the instrument), just as I described for the pushbutton Speak & Spell covered in Project 1 (Chapter 14).

Painting

All the same considerations apply as with the original Speak & Spell covered previously (Project 1, Chapter 14).

Control Mounting

Exactly as before, mount all components prior to soldering. As with all Speaks in the series from the membrane Speak & Spell onward, you have the two main choices of locating the group of three bending switches: either next to the speaker (refer to Figure 17-1) or along the side (Chapter 14, Figure 14-1).

Soldering

Using your bus wire, connect together the middle lugs of the three bending switches (refer to Figure 17-2). Be sure that the wire touches only the soldering lugs intended and nothing else.

Using your insulated wire-wrap wire, finish this three-switch matrix by soldering the indicated switch lugs to the circuit points shown. You'll be soldering four wires here: one to the three lugs you just connected together, and one to each of the three switch's lower lugs.

Next, wire the two switches of the looping matrix, exactly as shown. Wire the pitch dial as shown. Wire the body-contact as shown.

Wire the reset switch by snipping one of the wires coming from the battery compartment and soldering your normally closed pushbutton switch in the middle (see Appendix A for reset button wiring).

Note You might look for envelope LED bends as well. Several usually exist. Know, however, that some LED bends will keep the circuit from operating correctly depending on the version of the Speak & Math you're working on and where you wired the bend. If the circuit seems to stall or not start during the testing section that follows, remove the LED from the circuit and begin the test routine again. If the LED burns too brightly, add a 100 ohm resistor (brown-black-brown color code) to its anode (the *usually* longer lead and always the one terminating inside the LED as the smaller metal mass). Be sure to try different resistors when adjusting LED brightness.

Testing

So far, so good. Let's put this thing through its paces.

1. Turn the pitch dial all the way up and turn all your added toggle switches off.

2. Install fresh batteries.

3. Holding the batteries in, press the Speak's ON button. If all has gone well, you should hear the usual startup sequence. If not:

    a. Be sure that you've used the right switches in the right places (your N.O. and N.C. pushbuttons).

    b. Be sure that you've wired things correctly (the correct lugs on pot and toggle switches are connected as in the illustration).

    c. Be sure that the batteries are in correctly (two one way and two the other) and are making good contact.

Troubleshooting Tips

Don't be alarmed if a bent project doesn't work when it's first turned on. This happens a lot. Usually the problem is sitting right there in front of you, with just a little poking around needed to find it.

1. Hold the reset switch in as you start up. If this works, you've used a normally open instead of normally closed pushbutton. Install the correct switch.

2. Turn the pitch dial all the way down while you start up. If this works, you've wired the pot in reverse. Remove the wire from the outside lug and solder it to the other outside lug.

3. Do both of the preceding at the same time while you start up. If this works, install the correct pushbutton and reverse the wiring on the pot, as noted.

   If you're still having no luck, recheck all wiring for overdone solder joints or incorrect connections. When you hear the startup sequence and the instrument is operating normally, you can proceed with testing.

4. Test the reset switch. While the Speak is speaking, press the reset switch to confirm power supply interruption.

5. While the Speak & Math is in the middle of a routine (try all of its game functions), press and hold the pushbutton looping switch. If it's in the right routine, the sounds should begin to loop. Press and hold again for another loop, restarting a game routine if necessary should the Speak fall silent (or quickly hit the reset switch if the Speak crashes).

6. When you find a good loop, flip the looping toggle switch to lock the loop in place.

7. Try slowly turning the pitch dial down. Be careful when you see the display begin to shudder. This means that you're at the bottom end of the pitch range and turning the dial further will result in a crash.

Note See Chapter 9, Figure 9-23, as well as Appendix A, for instructions on adding a trimmer pot to the main pot to keep the main pot from crashing the circuit. You'll lose a little of the ultra-slow clocking range if you install a trim pot, but you'll never crash the bottom end of a loop again.

8. Last to test are the three bending switches. With pitch turned all the way up and the Speak doing one of its normal routines, try flipping one of the three bending switches. If nothing happens, try another of the Speak's routines. If on the right routine, the instrument should begin to produce streams of wild sound elements, some lasting a little while, some going on and on. These switches should all produce similar disruptions, singly or in combination.

Tip Remember to hit the reset switch, turn off the power switch, or quickly remove a battery upon the crashing of any bent instrument!

Reassembly

Same as before with all the models I've discussed. Just be sure to keep all wires from being pinched as you close the case. You knew I was going to say that.

Musicality

Because all Incantors output the same basic signals, musicality is, you'll remember, generally the same. Read the previous description of the original Speak & Spell's output for insights in Project 1 (Chapter 14). If you've built any of the previous models, I'm sure you'll notice immediately the improvement in voice fidelity. In addition, some of the Speak & Math's membrane keys will produce a stuttering effect when pressed. Experiment!

Project 5: The Casio SA-2 Aleatron

"A leatron" is the name I apply to circuit-bent keyboard instruments capable of producing aleatoric, or chance, music. Aleatrons can be absolutely fascinating in their strange, other-worldly, extra-musical output. This project, and the following five, supply you with a strong sampling of these alien music boxes in ascending order of difficulty. I end with the Casio SK-1, the classic sampler on everyone's bending list.

Though extremely simple, the SA-2 Aleatron (see Figure 18-1) is a favorite of mine and many other experimental musicians. Meant only to be a cool little voice and accompaniment keyboard, the nice sound quality (Pulse Code Modulation?), extravagant bent music, and very accessible circuitry make it a great target to bend.

What's so cool about the bent SA-2 Aleatron is how it produces its namesake—aleatoric, or chance, music. Unlike the Incantor's aleatoric music, an amalgam of allophones and sound effects, and unlike the SK-1's aleatoric music, waves of unidentifiable sounds and instruments, the SA-2's chance output is recognizable: often only drums, bass, and piano. It is the outlandish compositions rather than the outlandish sounds (which it also produces) that set this instrument apart. And it's this, listening to a drums/bass/piano jazz trio from another galaxy, that's so fascinating.

Line Out   Threshold Dial and Trigger   Body-Contacts

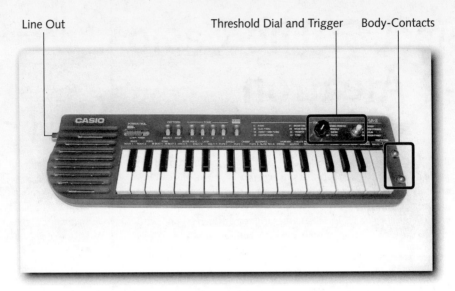

FIGURE **18-1: The SA-2 Aleatron with chance switches and body-contacts**

Parts

Before you begin this project, be sure to have all these parts at hand.

- 1 Casio SA-2 Keyboard
- 1 miniature N.O. (normally open) pushbutton switch
- 1 1M potentiometer (mini is best)
- 2 body-contacts
- Line-out array as desired (see Appendix A)

Open It Up!

Simple enough, until you get all the screws out and see that it's still not about to open. Why? For some mysterious reason, all along the front edge of the case there exists an interlock that must be pulled apart. Keeping the halves of the case flat to each other, begin at one end to pull the case halves apart. After a bad-news sound of plastic ripping, the halves do come apart without damage and the circuit is finally exposed.

Circuit at First Glance

Although we're usually interested in ICs, here you'll be looking mainly to connect a "one-shot" pushbutton to a couple of lone components further down the board (see bending diagram, Figure 18-2). It's interesting to note that although the battery compartment is located on the back half of the case, the battery contacts are located on the board right there in front of you. Keep this in mind as you run your wires. Just imagine the batteries lying between the battery contacts (the stiff wires rising from the board) and keep all wires from crossing this area.

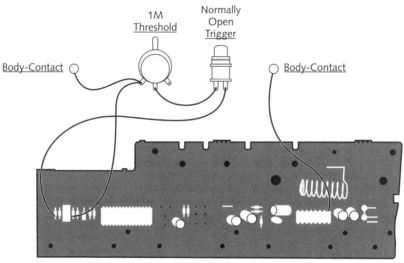

FIGURE 18-2: Bending diagram for the Casio SA-2 Aleatron. Capacitor C2's upper lead is the same as the IC pin that one of the body-contacts connects to (C2 is two capacitors to the right of the target IC pin). If C2's lead that is closest to the chromed battery spring is accessible, you can solder the body-contact wire there instead.

Step-by-Step Bending

You've gathered all the needed parts and you've opened things up. You've studied the diagram. Now, step by step, you'll complete the transformation. Here's how.

Choosing a Cool Control Layout

This time you have pretty much space to deal with as long as you're using small components. All components can be mounted in the area high and to the right of the front panel as in Figure 18-1 (behind this "Tone" area the circuit board is cut back, providing the needed space). You'll have a little room to deviate from this example if you want. Keep in mind not only ergonomics for playability but also mounting tool clearances while you plan your control configurations.

Because the line output will be derived from the speaker terminals, it makes sense to locate the output jack nearby. Before deciding upon a control layout, look at your line-out options in Appendix A. You'll need to add an output jack of some type, and you have the options of installing a speaker cut-off switch and even an envelope LED (if the speaker signal will drive it; it may not be strong enough; try various LEDs). If the line output is too strong for your needs, you'll want to include a trim pot in the scheme, also depicted in the appendix drawing.

Case Considerations

The back half of the case contains a series off reinforcing ribs, some of which might interfere with the new controls mounted opposite on the front panel. If so, just snip the ribs away where needed using a sturdy wire clipper. As to the front panel itself, there are really no concerns.

You'll note that it would be possible to mount a series of toggle switches within the exposed overhang visible at the back of the panel (turn the keyboard over and have a look). If you do this, you'll need to drill a hole in the back of the instrument, within this overhang, to accommodate your wire run. Look before you leap! The area above the speaker is easiest for this. But with care, you can also run your wires into the instrument through holes drilled carefully elsewhere within this trough.

Marking the Board

Looking closely at Figure 18-2, mark your board where all the new wires will attach. One wire will need to be soldered to an IC pin unless there's room to solder to the alternate location, a nearby capacitor lead (see the caption for Figure 18-2).

Drilling Holes

Mark all hole locations with your pencil. Using a $\frac{1}{8}$" drill bit, drill all your pilot holes. Open these holes to size with the hand reamer or, if the reamer's penetration is blocked, the careful use of a correctly sized burr bit mounted in your Dremel drill. Finish the hole edges with your de-burrer.

Painting

Removal of the circuit board and speaker is easy. If you're intending to paint the instrument and want to keep the keyboard unpainted (or want to paint it differently), removal is mandatory. After the circuit and speaker are removed, the keyboard will lift out without a problem. Handle the keyboard with care. It's quite fragile.

Control Mounting

Fairly straightforward. Try to mount your RCA output jack, allowing you room for tightening the hex nut within the speaker enclosure without too much difficulty. If you're mounting the body-contacts as shown in the illustration, you might find it helpful to remove the circuit board to better access the bolt heads, though this probably won't be necessary.

Soldering

After all components are mounted, simply solder things together as shown in Figure 18-2. When soldering to the tiny resistors to the left of the board, you it might find it helpful to "play hooky" as covered in Chapter 8. In brief, form a small hook in the end of your stripped wire-wrap wire and hook the target resistor lead firmly. With a very clean soldering tip, finish the job.

Line-output circuitry is the same in all projects if not otherwise noted. See the line-output scheme in Appendix A and assume that the speaker in the diagram is the speaker in the project.

Testing

Because the battery compartment cannot be connected to the instrument in any way other than by bringing the two halves of the case together, do just that. Replace a couple screws to keep the halves from coming apart. Insert fresh batteries.

Turn the SA-2 Aleatron on and hit the demo tune (I hope you like it; you'll be hearing it a lot). With the chance dial turned down, hit the chance trigger (pushbutton) while the demo is running. Nothing should happen. If the instrument's audio is interrupted, you've wired the pot in reverse. Open the case and de-solder the wire going to the outside lug of the pot. Solder it to the opposite outside lug instead.

Now, with the demo tune running, hit the chance trigger as though you're tapping a bongo drum (in other words, don't push it in like a doorbell). *Tapping* is key here.

Nothing should happen until you get the dial turned up high enough. When the dial approaches the trigger threshold the music will begin to be interrupted. The further you turn the dial, the more "kick" will be delivered through the trigger button. But you don't need much. For now, leave the dial set to where it just begins to influence the trigger.

The keyboard will probably crash. A lot. Each time it does, turn it off and back on. Begin the demo tune again. Sooner or later the trigger will initiate a passage of aleatoric, or chance, music. The dial may now be backed off a little. Further quick tapping of the trigger will augment the chance phases of music.

When a chance passage is happening, try touching the chance body-contacts. This system has a slighter effect upon the chance phrases, incrementing them forward more delicately than possible with the mechanical trigger. Which musician on stage is the circuit-bender? That's easy. It's the one licking his or her fingers. If body-contacts ever seem insensitive, there's the popular solution (but don't expect to be invited to your community black-tie affairs).

Note It's possible to wire a reset switch to this circuit to allow a quicker reset than by using the original slide power switch. This is done by cutting a power supply trace on the back of the circuit board (see Chapter 9 for complete instructions). Simply observe where the stiff battery contact wire connects to a circuit trace and cut the trace as detailed in Chapter 9.

Reassembly

You already have the case halves together. Check to be sure that no wires are running across the area where the batteries are housed, and that none are in the way of the key mechanism. If all's well, finally replace the rest of the screws and you're done.

Musicality

I enjoy clipping bits of the chance phrases and turning these into loops that I can accompany with lead voices or background elements. But even more, I simply enjoy the SA-2 Aleatron as an outrageous music box.

Though I've been listening for years, the SA-2 will still surprise me with new output. It will you, too. Don't let the crashes set you back. Keep at it. There will be fuzzy phrases, dirty noise fests, and some audacious cacophony. But, in time, there will also be crystal-clear romps of eccentric composition, unlike anything you've ever heard. Not the contrived, made-for-Hollywood *Star Wars* cantina tunes played by sticky puppets, but real alien music, produced by you.

Project 6: The Casio SA-5 Aleatron

If you're not content waiting to find the (rather rare) Casio SA-2 to carve your own Aleatron out of, be calmed. There are two other small keyboards that can be bent in just about the same way. These are the Casio SA-5 and the Realistic Concertmate 380. Among these three, sooner or later you'll find one to bend. Take a look at the SA-2's introduction in the previous project for more on Aleatrons and chance music.

Parts

Before you begin this project, be sure to have all these parts at hand.

- 1 Casio SA-5 Keyboard
- 1 miniature N.O. (normally open) pushbutton switch
- 1 1M potentiometer (mini is best)
- 2 body-contacts
- Line-out array as desired (see Appendix A)

Open It Up!

The SA-5 opens exactly the same as the SA-2 described in the prior project. Take a look back if it's unfamiliar.

Circuit at First Glance

Although the circuit doesn't look much like the circuit of the SA-2's, you'll bend it in primarily the same way (see Figure 19-2). One big difference, however, is that you'll need to do a little soldering to the large IC at the left of the board.

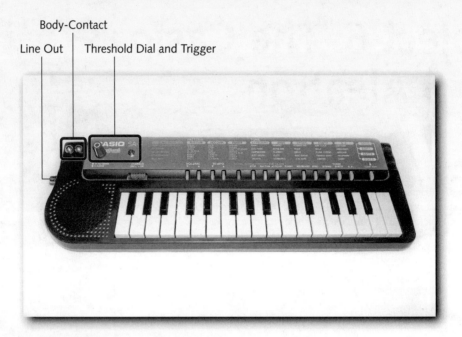

FIGURE **19-1: The Casio SA-5 Aleatron**

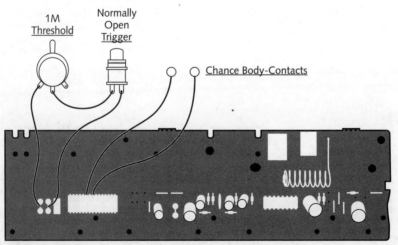

FIGURE **19-2: Bending diagram for the Casio SA-5 Aleatron**

Step-by-Step Bending

You've gathered all the needed parts and you've opened things up. You've studied the diagram. Now, step by step, you'll complete the transformation. Here's how.

Choosing a Cool Control Layout

Unlike with the SA-2, you'll not find much room within the SA-5 to work with. There's the usual cramped space around the speaker where you'll again mount the RCA output jack. And there's the small strip of space to the right of the highest key on the keyboard. The other available space is the vast panel above the colorful pushbuttons, where you'll find more room than needed to mount your controls.

Because the line output will be derived from the speaker terminals, it makes sense to locate the output jack nearby. Before deciding upon a control layout, look at your line-out options in Appendix A. You'll need to add an output jack of some type, and you have the options of installing a speaker cut-off switch and even an envelope LED (if the speaker signal will drive it; it may not be strong enough—try various LEDs). If the line output is too strong, you'll want to include a trim pot in the scheme, also depicted in the drawing.

Case Considerations

As you plan your control locations, keep in mind that the higher up the back panel you mount your controls, the more likely the instrument is to tip backward as you use them. Mounting the controls low on the panel makes sense.

When mounting controls on the panel above the colored buttons, you'll need to drill a hole to run your wires through into the inside of the instrument. To avoid drilling (or running wires) near the edge of the circuit board, I drill a small hole (⅛") in the wall just above the speaker and do my wire run right through it, past the speaker and on to the circuit.

Marking the Board

Following the bending diagram (refer to Figure 19-2), mark all locations you'll be soldering wires to (the wires that go to the IC connect to pins 6 and 8, counting over from the left). This time the two wires that go to the chance dial and trigger are soldered to the top wire on two arch-like components, making this connection a little easier than on the Casio SA-2.

Drilling Holes

Mark all hole locations with your pencil. Using a ⅛" drill bit, drill all your pilot holes. Open these holes to size with the hand reamer or, if the reamer's penetration is blocked, the careful use of a correctly sized burr bit mounted in your Dremel drill. Finish holes with your de-burrer.

Painting

As with the SA-2, removal of the circuit board and speaker is easy. If you're intending to paint the instrument and want to keep the keyboard unpainted (or want to paint it differently), removal is mandatory. After the circuit and speaker are removed, the keyboard will lift out without a problem. Handle the keyboard with care. It's quite fragile.

Control Mounting

Fairly straightforward. Try to mount your RCA output jack, allowing you room for tightening the hex nut within the speaker enclosure without too much difficulty. If you're mounting the body-contacts as in the illustration, you might find it helpful to remove the circuit board to better access the bolt heads, though this probably won't be necessary.

Soldering

After all components are mounted, simply solder things together as shown in Figure 19-2. When you're soldering to the arched components to the left of the board, you might find it helpful to "play hooky" as covered in Chapter 8. Form a small hook in the end of your stripped wire-wrap wire and hook the target lead firmly. With a very clean soldering tip, finish the job. Make this connection as quickly as possible.

Line-output circuitry is the same in all projects if not otherwise noted. See the line-output scheme in Appendix A and assume that the speaker in the diagram is the speaker in the project.

Testing

Because the battery compartment cannot be connected to the instrument in any way other than by bringing the two halves of the case together, do just that. Replace a couple of screws to keep the halves from coming apart. Insert fresh batteries.

Turn the SA-5 Aleatron on and hit one of the demo tunes. With the chance dial turned down, hit the chance trigger (pushbutton) while the demo is running. Nothing should happen. If the instrument's audio is interrupted, you've wired the pot in reverse. Open the case and de-solder the wire going to the outside lug of the pot. Solder it to the opposite outside lug instead.

Now, with the demo tune running, hit the chance trigger as though you're tapping a bongo drum (in other words, as with the SA-2 Aleatron covered prior, don't push it in like a doorbell). *Tapping* is key here.

Nothing should happen until you get the dial turned up high enough. When the dial approaches the trigger threshold, the music will begin to be interrupted. The further you turn the dial, the more "kick" will be delivered to the trigger button. But you don't need much. For now, leave the dial set to where it just begins to influence the trigger.

The keyboard will probably crash a lot. Each time it does, turn it off and back on. Sooner or later the trigger will initiate a passage of aleatoric, or chance, music. The dial may now be backed off a little. Further quick tapping of the trigger will augment the chance phases of music.

When a chance passage is happening, try touching the chance body-contacts. This system has a milder effect upon changing the chance phrases, incrementing them forward more delicately than is possible with the mechanical trigger. If you don't mind being snubbed by high society, finger-licking will increase the conductivity of your fingertips if your skin is dry, or when playing in low-humidity atmospheres.

Note It's possible to wire a reset switch to this circuit to allow a quicker reset than by using the original slide power switch. This is done by cutting a power supply trace on the back of the circuit board (see Chapter 9 for complete instructions). Simply observe where the stiff battery contact wire connects to a circuit trace and cut the trace as detailed in Chapter 9.

Reassembly

You already have the case halves together. Check to be sure that no wires are running across the area where the batteries are housed, and that none are in the way of the key mechanism. If all's well, finally replace the rest of the screws and you're done.

Musicality

The SA-5 Aleatron's chance output routines are very similar to that of the SA-2's covered in the previous project. Take a look back.

Project 7: The Realistic Concertmate 380 Aleatron

Once again, here you have a wonderful target for creating a powerful Aleatron. The instructions are essentially the same as those for the SA-5, just covered. In fact, other than the case graphics, the Concertmate 380 and the Casio SA-5 are *exactly the same*. Looks as though anyone can slap his or her name on the thing! Radio Shack's version of the Casio instrument is quite common, appearing in the thrift shops frequently. See Figure 19-1 (Chapter 19) for the SA-5's control layout; the 380's is identical.

Parts

Before you begin this project, be sure to have all these parts at hand.

- 1 Realistic Concertmate 380 keyboard

- 1 miniature N.O. (normally open) pushbutton switch

- 1 1M potentiometer (mini is best)

- 2 body-contacts

- Line-out array as desired (see Appendix A)

Open It Up!

The Concertmate 380 opens exactly the same as the SA-5 and the SA-2 described in the prior projects. Take a look at the SA-2 for details.

Circuit at First Glance

The circuit looks exactly like the SA-5's, and you'll bend it in exactly the same way (see Figure 20-1).

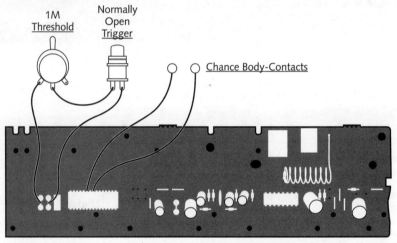

FIGURE 20-1: Bending diagram for the Radio Shack Concertmate 380

Step-by-Step Bending

You've gathered all the needed parts and you've opened things up. You've studied the diagram. Now, step by step, you'll complete the transformation. Here's how.

Choosing a Cool Control Layout

Unlike the SA-2, you'll not find much room within the Concertmate 380 to work with. There's the usual cramped space around the speaker where you'll again mount the RCA output jack. And there's the small strip of space to the right of the highest key on the keyboard. The other available space, as in the 380's near-twin, the Casio SA-5, is again within the vast panel above the colorful pushbuttons.

Because the line output will be derived from the speaker terminals, it makes sense to locate the output jack nearby. Before deciding upon a control layout, look at your line-out options in Appendix A. You'll need to add an output jack of some type, and you have the options of installing a speaker cut-off switch and even an envelope LED (if the speaker signal will drive it, it may not be strong enough). If the line output is too strong, you'll want to include a trim pot in the scheme, also depicted in the drawing.

Case Considerations

As you plan your control locations, keep in mind that the higher up the back panel you mount your controls, the more likely the instrument is to tip backward as you use them. As in the example of the SA-5, mounting the controls low on the panel makes sense.

When mounting controls on the panel above the colored buttons, you'll need to drill a hole to run your wires through into the inside of the instrument. To avoid drilling (or running wires) near the edge of the circuit board, I drill a small hole ($^1/_8$") in the wall just above the speaker and do my wire run right through it, past the speaker and on to the circuit.

Marking the Board

Following the bending diagram (refer to Figure 20-1), mark all locations you'll be soldering wires to (the wires that go to the IC connect to pins 6 and 8, counting in from the left). This time, the two wires that go to the chance dial and trigger are soldered to the top wire on two arch-like components, making this connection a little easier than on the SA-2.

Drilling Holes

Mark all hole locations with your pencil. Using a $^1/_8$" drill bit, drill all your pilot holes. Open these holes to size with the hand reamer or, if the reamer's penetration is blocked, the careful use of a correctly sized burr bit mounted in your Dremel drill. Finish hole edges with your de-burrer.

Painting

As with the SA-2, removal of the circuit board and speaker is easy. If you're intending to paint the instrument and want to keep the keyboard unpainted (or want to paint it differently), removal is mandatory. After the circuit and speaker are removed, the keyboard will lift out without a problem. Handle the keyboard with care. It's quite fragile.

Control Mounting

Fairly straightforward. Try to mount your RCA output jack, allowing you room for tightening the hex nut within the speaker enclosure without too much difficulty. If mounting the body-contacts as in the illustration (Figure 19-1, one project back), you might find it helpful to remove the circuit board to better access the bolt heads, though this probably won't be necessary.

Soldering

After all components are mounted, simply solder things together as seen in Figure 20-1. When soldering to the arched components to the left of the board, you might find it helpful to "play hooky" as covered in Chapter 8. Form a small hook in the end of your stripped wire-wrap wire and hook the target resistor firmly. With a very clean soldering tip, finish the job. Make this connection as quickly as possible.

Line-output circuitry is the same in all projects if not otherwise noted. See the line-output scheme in Appendix A and assume that the speaker in the diagram is the speaker in the project.

Testing

Because the battery compartment cannot be connected to the instrument in any way other than by bringing the two halves of the case together, do just that. Replace a couple screws to keep the halves from coming apart. Insert fresh batteries.

Turn the Concertmate 380 Aleatron on and hit one of the demo tunes. With the chance dial turned down, hit the chance trigger (pushbutton) while the demo is running. Nothing should happen. If the instrument's audio is interrupted, you've wired the pot in reverse. Open the case and de-solder the wire going to the outside lug of the pot. Solder it to the opposite outside lug instead.

Now, with the demo tune running, hit the chance trigger as though you're tapping a bongo drum (as before, don't push it in like a doorbell). *Tapping* is key here.

Nothing should happen until you get the dial turned up high enough. When the dial approaches the trigger threshold, the music will begin to be interrupted. The further you turn the dial, the more "kick" will be delivered to the trigger button. But you don't need much. For now, leave the dial set to where it just begins to influence the trigger.

The keyboard will probably crash, a lot. Each time it does, turn it off and back on. Sooner or later the trigger will initiate a passage of aleatoric, or chance, music. The dial may now be backed off a little. Further quick tapping of the trigger will augment the chance phases of music.

When a chance passage is happening, try touching the chance body-contacts. This system has a milder effect upon changing the chance phrases, incrementing them forward more delicately than possible with the mechanical trigger. As you know by now, finger licking is allowed, as long as the fingers are your own.

 Note It's possible to wire a reset switch to this circuit to allow a quicker reset than by using the original slide power switch. This is done by cutting a power supply trace on the back of the circuit board (see Chapter 9 for complete instructions). Simply observe where the stiff battery contact wire connects to a circuit trace and cut the trace as detailed in Chapter 9.

Reassembly

You already have the case halves together. Check to be sure that no wires are running across the area where the batteries are housed, and that none are in the way of the key mechanism. If all's well, finally replace the rest of the screws and you're done.

Musicality

The Concertmate 380's chance output routines are very similar to that of the SA-2's, covered two projects back. Take a look at my thoughts, but more important, listen for yourself and face the oft-encountered dilemma of experimental music: dealing with chance.

Project 8: The Casio SK-60 Aleatron

Casio's success story with the SK-1, the world's first sampler for the masses, was encouraging enough for the company to launch further "SKs" (**S**ampling **K**eyboards) as the years rolled on. Sample time as well as the number of user samples increased, with the SA-60 now allowing four sounds to be sampled as compared with the Casio SK-1's single sample. Thinking, probably correctly, that the most frequently sampled sound is the user's voice, this voice-as-sample theme was fully embraced when Casio released the quite odd SK-60.

Whereas most musical keyboards tend to focus upon an array of musical instrument voices held within their sound banks, Casio's SK-60 takes another direction. Along with the prerequisite piano, strings, and trumpet come all kinds of human voices. Eighteen varieties of human voice, in fact, ranging from children to adult, male and female, and singing everything from "ahhh" to "do-bee-do-bah." Naturally, this thing needs to be bent (see Figure 21-1).

Parts

Before you begin this project, be sure to have all these parts at hand.

- 1 Casio SK-60 Keyboard
- 10 miniature SPST (or SPDT) toggle switches
- 1 1M potentiometer
- 1 miniature N.O. (normally open) pushbutton switch
- Line-out array as desired (see Appendix A)

Flux Dial and Reset on Side

LED   Line Out

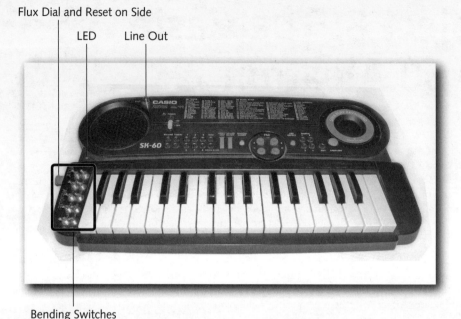

Bending Switches

FIGURE 21-1: The Casio SK-60 Aleatron

Open It Up!

Although the case design is somewhat convoluted, you'll still have good access to the circuit when the back is opened and your wrist recovers from removing all those screws. The metal of the screws themselves is rather soft, and the slots in the screw heads are shallow. This is a bad combination.

Why? Well, if the Phillips driver you use to remove the screws isn't a good fit, there's the concern of damaging the slots, making the screws' removal much less fun. Or impossible. Any time you run into such screws, match them with a driver whose fit is deep and snug. Too small or too large a driver bit can result in real trouble here. No matter what screw you're working on, it's a good idea to *always* use the correct driver size even if the mismatched driver closer at hand can be made to work.

After all the screws are removed (and set aside in a safe place, such as on your parts magnet), you're ready to separate the case halves.

The back section will remain in one large piece, including the entire keyboard. Removing the screws allows you only to separate the front panel from the rest of the assembly.

The next step is to de-solder the speaker wires so that you'll be able to remove the entire front panel from the instrument. Note that the red wire goes to the positive terminal of the speaker, the white to negative.

Set the front panel aside and out of the way. Be careful with this section because it contains the upper half of the sliding power switch (note the shiny metal tongues) as well as the rubber sheets that locate the additional pushbutton switch contacts. Both these components can fall out of their places; note their original orientation in case you find yourself trying later to figure out where these pancakes went!

Circuit at First Glance

The actual IC looks a little intimidating until you realize that all the tiny leads you need to work with extend out into much larger traces. Bending the SK-60 is a matter of getting all the larger traces right and in order—easy if you follow the bending diagram carefully (see Figure 21-2).

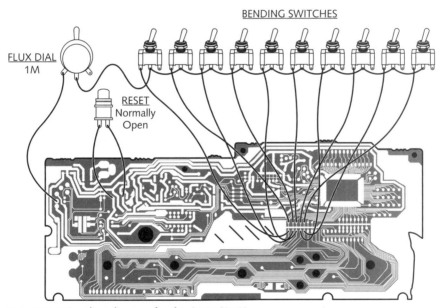

FIGURE 21-2: Bending diagram for the Casio SK-60

Note the dull black traces that serve as the contact areas for the front panel switches. Keep in mind that any wires you solder to the board must be run to their components while entirely avoiding the dull, black traces. If not, the newly installed wires could easily interfere with the operation of the switches.

Step-by-Step Bending

You've gathered all the needed parts and you've opened things up. You've studied the diagram. Now, step by step, you'll complete the transformation. Here's how.

Choosing a Cool Control Layout

As shown in Figure 21-1, I've placed the bank of bending switches in the open wing to the left of the keyboard. A similar area to the right of the keyboard also avails itself of the opportunity for additions, as do a small area around the speaker and an area to the right and above the microphone.

Space is somewhat tight if you intend to mount all ten switches as I have, in one wing. Splitting these ten switches into two groups is another consideration. The first five could go in one wing, the remaining five in the other wing. Such an orientation, if planned well, will also allow more room for your side-mounted 1M pot (especially if you use a full-sized pot instead of the miniature pot shown in Figure 21-1).

Because the line output will be derived from the speaker terminals, it makes sense to locate the output jack nearby. Before deciding upon a control layout, look at your line-out options in Appendix A. You'll need to add an output jack of some type, and you have the options of installing a speaker cut-off switch and even an envelope LED (if the speaker signal will drive it; it may not be strong enough—try various LEDs). If the line output is too strong, you'll want to include a trim pot in the scheme, also depicted in the appendix drawing.

Case Considerations

The SK-60 is made of friendly plastic—"friendly," that is, should workability rather than land-fill half-life be the point. Beyond this, there are no ribs in the way, and the only case alterations beyond component mounting will be the hole you'll need to drill to run your wires from the wing into the main body of the instrument (again, see Figure 21-1).

Marking the Board

Following the bending diagram (refer to Figure 21-2), mark the board to show all points you'll be soldering wires to. Yes, I know, sometimes this seems redundant. In the case of unmistakable points, sure, it's okay to just look at a wiring diagram and solder away. But I've always found that ascertaining connection points prior to the moment of soldering keeps things cheerful.

Drilling Holes

Using a sharp pencil, and after considering both component and mounting tool clearances, mark all component locations. Next step is to drill all the pilot holes with your ⅛" bit. Bring these holes up to component size with your hand bore. Finish the holes with your de-burrer.

Here, and in other projects, you'll find that the de-burrer bit doesn't quite reach into the tight areas you'd like it to. In such a case, plastic burrs can also be removed with the blade of a small knife or screwdriver, or even a spinning burr bit mounted in your Dremel drill.

Painting

With a little disassembly, painting the SK-60 is easy. It should be painted in two sections, with the body and front panel each having its own painting session.

From the front panel you'll need to remove all the switch components. The rubber contact sheets will drop right off their posts. The contact section of the power switch will be freed when the slider handle on the other side of the panel is pried away from it.

As mentioned previously, notice the orientation of these components before they're removed from the panel. At times I'll mark such components with registration lines to be certain that I'm aligning everything correctly upon reassembly.

You have two ways to go on the circuit and keyboard: mask or removal. By all means, removal is the best way to go.

Remove the retaining strip at the top of the keyboard keys. Next, remove the screws holding the circuit board in place. You'll also need to remove the small circuit board next to the microphone (and try to guess what's on the other side before you remove it—did you guess right?).

Carefully remove the keys and set aside in a very secure spot. Beneath the keys you'll find the switch panel that the keys hit when pressed. The soft rubber strip should stay in place as you remove the final screws holding the panel to the bottom of the instrument's case.

Last, lift the microphone out of its nest, trying your best not to disturb the foam it's wrapped in.

At this point you should be able to remove the electronics from the case and proceed to painting, following the instructions in Chapter 12.

If you decided to mask instead of taking everything apart, well, okay. It's possible. You'll still need to remove all the switch components from the top panel. After that, use masking tape to mask the keyboard. To do this correctly you'll need to insert tape into the thin gaps between the keyboard and the surrounding case. Fold the remaining tape over the keys and press down. Mask the remaining key area as well as the circuit board itself.

You can now proceed with painting following the procedures, again, in Chapter 12.

The chief problem with masking instead of removal of the keyboard is that the tape-to-case gaps around the keyboard can cause paint to gather and fill these small spaces. If this happens, the paint job will be ruined where the tape has caused interference. Stripping a case of any and all components prior to painting is *always* the better solution.

Note Anytime you paint over a panel containing complex control legends, you might want first to scan the panel for future reference or retitling, unless you're the type who revels in confusion. Like me.

After the paints have dried, you can replace all the components you removed prior to painting and proceed to the next section.

Control Mounting

If you're following the control layout in Figure 21-1, control mounting should be a snap. The only point to bring forward is the usual consideration when mounting bare LEDs: Use the correct bit size and drill from the inside out. If you're placing the LED inside a pilot light assembly, hole size and finish will, of course, not be as critical.

Soldering

This time out, the most challenging soldering task will be making connections to the ten points just below the IC. But because you're using thin wire-wrap wire, all will go as smoothly as possible.

Be accurate and steady handed; use as little solder as needed to create small, smooth connections.

As you explore certain circuits, you'll discover that shorting two points to each other will reset the circuit you've just crashed with your coolest bend. Rather than the usual power-supply interrupt to implement a reset, as most projects reflect, this secondary reset process, although unusual, is also valuable. For the situation in which power-supply interruption is difficult to implement, this short-circuit reset becomes invaluable. But it often comes at a price.

The price of the short-circuit reset switch is that it might be dangerous to use, depending on how you use it. And such is the case with the reset on the SK-60. Instead of the usual N.C. (normally closed) pushbutton switch used in the power supply interrupt, the short-circuit reset uses an N.O. (normally open) pushbutton. Pressing the button for a split second sends a small zap through the circuit and, like a cardiac defibrillator applied to a heart attack victim, starts things up again.

To extend the defibrillator example, keeping the lightning bolts flowing into the newly resuscitated patient for, say, an hour or two, would be ill advised and probably the most exciting topic in the ensuing wrongful death suit. Get the picture? You're soldering a bent defibrillator into the SK-60. Tap it as briefly as humanly possible and as if the instrument's life is depending upon your doing so. That's because it does.

Too scary? Remember, in the event of a deep crash, when turning the power switch on and off won't correct the situation, you can always break contact within the battery compartment to reset the circuit (dislodge a battery for a moment).

Line-output circuitry is the same in all projects if not otherwise noted. See the line-output scheme in Appendix A and assume that the speaker in the diagram is the speaker in the project.

Testing

After reconnecting the speaker to the circuit, you can close the case to proceed with testing. As you close the case, be sure that the rubber switch sheets and the power switch contacts are positioned correctly. Leave the screws out for now, unless you want to replace a couple just to keep things closed while you do the following tests.

I already touched upon the thing to be aware of in closing the case: your wire runs. Just be sure that the new wires are away from the dull black traces on the circuit board. Again, these are the conductors that the mechanical switches on the control panel must touch to operate correctly. Blocking any of these black traces with your wires will interfere with their operation.

Be sure that the pot is turned down all the way and the new switches are all turned off.

Remember the defibrillator you installed? Remember to only tap it? And that holding it down can fry the circuit? Good. Give it a tap while the demo tune is running. If you've wired it correctly, the demo should stop and the instrument should be reset. Again—only tap this switch as briefly as you can. *Never* press it like a doorbell because, trust me, no one will be home.

If you've installed an LED, it should be flashing along with the music by now (see Appendix A, line-output scheme). If not, reverse the leads and try again. If still no luck, try another LED.

Continue testing by recording four samples, one into each sample bank (four different words spoken into the microphone will be fine). Using the Edit function, have these four samples play as a repeating loop.

Turn one of the bending switches on. Nothing should happen yet. If there is a response it's probably because the pot has been wired backwards. If so, just switch the wire going to the outside lug to the other outside lug.

With your loop of four samples running and one of the bending switches turned on, slowly begin to turn the pot up. At a certain point the loop will explode into a loud chord effect. When this happens, slowly back off the pot until just before the chord erupts.

What you're looking for here is the very slim margin between normal play and the loud chord sound. It's hard to find. Turn the pot back and forth until you find it. When you get the pot set just right, your sample loop will reform into an assortment of unusual compositional structures.

You'll find that the slim "chance zone" can also be accessed by changing the main volume setting when the pot is tuned in the correct range. You can use the volume control as a fine-tuning function in this regard. Experiment with the tempo control, too. You'll find it capable of changing the patterns of the bent music in play.

Now go deeper. Experiment with demo tunes, various voices, keyboard interactions, and whatever else you might be able to introduce the bending switches into. Combine switches. Reset the pot. Adjust the tempo again. See what else you can find! There's a lot to discover within the bent SK-60.

Reassembly

Because you've already closed the case and are certain that your wire runs are out of the way, you're just about finished. Replace the remaining screws, being careful that no wires are blocking the screw holes.

Musicality

Although playing the keyboard *is* possible within some of the bent routines (don't expect it to respond as usual), I look at this aleatoric music generator the way I see most: as alien chance music boxes. But that's not to say it all ends there. Recording these chance passages as a basis for further composition is often very rewarding.

Listening to abstract music is like taking an inkblot test: What does the blot look like? In considering what experimental sound "looks" like, experimental composers might be inspired to fill out these abstract musical forms the way the traditional composer orchestrates musical ideas and produces variations upon a theme.

Project 9: Casio MT-140 Aleatron

"Fairly unspectacular" might be an appropriate description for the Casiotone MT-140, at least in its pre-bent, sales floor condition. But that's exactly why I include it here. The MT-140 is so representative of countless other midsize-key keyboard instruments, stranded somewhere between "toy" and "serious" in their design, that it will serve as a perfect example of this otherwise forgettable genre (see Figure 22-1).

But once bent, the MT-140 steps outside of its wallflower self and becomes a seriously disturbed, hard-to-ignore individual. In bending, that's a good thing.

Parts

- 1 Casio MT-140 Keyboard
- 17 miniature SPST (or SPDT) toggle switches
- 1 1K potentiometer
- 1 body-contact
- line-out array as desired (see Appendix A)

Open It Up!

Simply remove the series of screws on the bottom of the case and it will open easily. Don't let the "screw" in the central hole fool you. It does look like a Phillips head screw down there, and were you to miss the "tune" designation, you might be tempted to apply the screwdriver. This is actually the dial of a trimmer pot made accessible here, as on many keyboards, for tuning the entire keyboard to the correct pitch. Leave it alone.

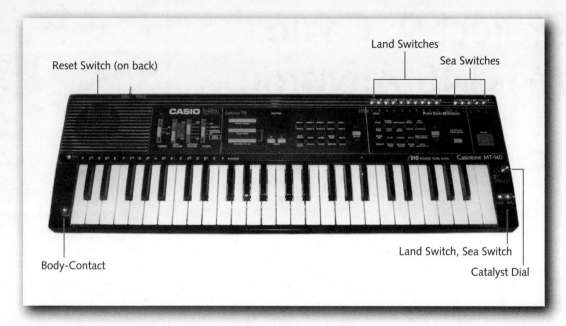

FIGURE 22-1: The Casio MT-140 Aleatron

Be aware that the battery compartment wires connect both halves of the case together. Observe where they connect to the circuit, and note polarity so that you'll be able to reconnect them correctly during final reassembly. De-solder these wires and place the bottom half of the case in a safe place.

Note that when you re-solder these wires to the circuit board, you'll really be re-soldering only one to its original position. The other wire will run through an N.C. (normally closed) push-button switch before connecting to the circuit board, with the switch serving as a power-supply-interrupt reset switch.

Circuit at First Glance

Anytime people see a large IC radiating a mass of near-microscopic printed circuit traces, a certain sense of apprehension commonly flows over the layperson. But you're a bender. Be happy. This usually means that there's lots to discover. And as we always hope, the tiny traces on the MT-140 lead to larger areas, making soldering a doable task (see bending diagram, Figure 22-2).

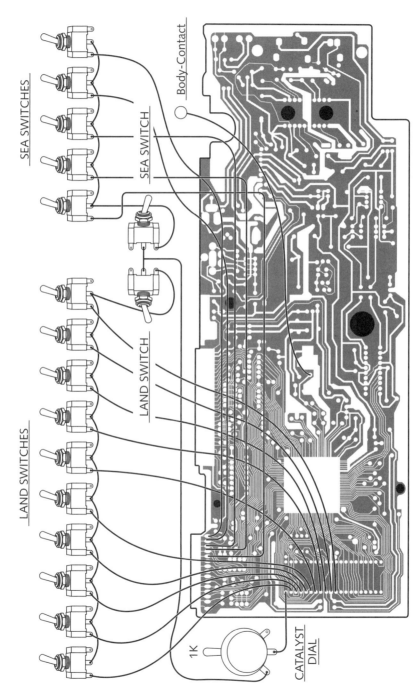

SEA SWITCHES

SEA SWITCH

Body-Contact

LAND SWITCHES

LAND SWITCH

1K

CATALYST
DIAL

Figure 22-2: Bending diagram for the Casio MT-140 Aleatron

Step-by-Step Bending

You've gathered all the needed parts and you've opened things up. You've studied the diagram. Now, step by step, you'll complete the transformation. Here's how.

Choosing a Cool Control Layout

Figure 22-1 shows the line of 15 bending switches placed where the case of the instrument best allows them. You'll find much more space available below the group of five switches, at the sides of the keyboard, and along the sides of the case in certain places.

There are *many* more bends possible on the MT-140 than I've implemented in the project model. If you further search the rows of pins adjacent to the pins designated as soldering points in the bending diagram (refer to Figure 22-2), I'm sure you'll find additional strong bends. If you want to expand upon this project, consider using sub-miniature switches, the next size down from those used in these projects. Using sub-miniature toggle switches will allow more switch-per-inch arrangement, and in this sense will nearly double your control-mounting space.

The patch bay (Appendix A) also answers this need and can be configured in countless ways. Imagine an instrument absolutely covered with pin jacks (the next size down from banana jacks), cross-connected everywhere with patch cords. The MT 140 will probably support such a project.

Because the line output will be derived from the speaker terminals, it makes sense to locate the output jack nearby. Before deciding upon a control layout, look at your line-out options in Appendix A. You'll need to add an output jack of some type, and you have the options of installing a speaker cut-off switch and even an envelope LED (if the speaker signal will drive it; it may not be strong enough—try various LEDs). If the line output is too strong, you'll want to include a trim pot in the scheme, also depicted in the appendix drawing.

Case Considerations

To me, the plastic used in this case seems a little thinner and more brittle than it should for an instrument of this size. This should raise a flag in drill-happy benders. The "proceed with caution" warning is due. General drilling procedures should be sharpened, as covered in a moment.

Marking the Board

Mark, as usual, the soldering points as indicated in the bending diagram (refer to Figure 9-2). They'll be easy to keep track of because each of the two sets aligns without breaks, each in a single row. Note that the pot connects to the top pin of the row that the ten Land switches connect to.

Drilling Holes

When you're sure of component placement, having checked for component as well as mounting tool clearances, mark the hole positions with a sharp pencil. Use the $\frac{1}{8}$" bit to drill all your pilot holes. If you're planning to include an LED in your line-output scheme (see Appendix A), remember to drill the correct-sized hole from the inside out, inserting the LED in the same direction the drill bit entered the plastic. You'll remember that the "exit" side of a drilled hole is cleaner than the "entrance" side and will look better around the bare LED.

As mentioned previously, the plastic of the MT-140's case seems a little fragile. A little too thin. But if you go slowly with your drilling, using a medium-fast drill speed and not applying much pressure upon the plastic (let the drill do the work), all should go well.

Try to keep the drill bit at 90 degrees to the plastic throughout the drilling procedure (unless, of course, the hole is meant to penetrate at another angle for some reason). The point is that when drilling thin or fragile materials at whatever angle, maintaining a consistent alignment throughout the drilling procedure becomes even more important.

When all holes are brought up to size, finish all hole edges (except LED hole) with your de-burrer.

Painting

Rather than detail the lengthy procedure needed to disassemble the innards of the MT-140, let me just say that there are no surprises. The disassembly is simply a common-sense procedure. Keep track of all parts removal sequences as well as all parts and their original placement (there will be lots of loose parts after disassembly, including all the buttons from the front panel).

When the case is free of all circuits, buttons, and the keyboard, painting can proceed (see Chapter 12).

Control Mounting

As long as your holes have been located correctly, control mounting should present no special problems on the MT-140. If you find that the potentiometer runs into the black plastic wall at the keyboard's edge, don't worry. Enlarge the hole a little, re-center the pot, and tighten the hex nut.

Soldering

If you remember the "playing hooky" soldering technique (Chapter 8), well, good for you. Now's the time to use it. Don't recall? Refresh your memory by going back and perusing the subject once more.

Playing hooky with the wires needing to be soldered to rows of IC pins is a fine technique, especially here. You can even wrap the other end of each wire around a screw post off to the side of the circuit to keep the wire in place for soldering as you go, one wire after the other until all are connected to the board.

Remember that "fast and precise" connections are mandatory when soldering to IC pins so as to keep heat down and solder accumulation to a minimum. And use only tiny bits of solder.

Connecting a reset switch to the power supply is simple. Just interrupt either wire with an N.C. (normally closed) pushbutton switch.

Line-output circuitry is the same in all projects if not otherwise noted. See the line-output scheme in Appendix A and assume that the speaker in the diagram is the speaker in the project.

Testing

After all wiring is complete, including reconnecting the battery compartment to the circuit board through the reset switch, testing can commence. Be sure that all new switches are turned off and the pot is turned down.

Start a rhythm and hit the reset switch. The circuit should reset. If the instrument won't start unless the reset switch is held in, you've mistakenly installed an N.O. (normally open) instead of an N.C. (normally closed) pushbutton switch for your reset. Install the correct switch and proceed.

If you've installed an envelope LED within the line output circuit, you should see it flashing at volume peaks. If not, switch its leads around to reverse its polarity. If still no response, try other LEDs. A "hi-brightness" (over 2,000 mcd) LED of a 3–4 volt rating should work well.

With a rhythm playing, turn on one Land switch. Nothing should happen until you turn up the Catalyst pot. If the rhythm is interrupted as soon as you turn the Land switch on with the pot turned all the way down, turn it all the way up and start again. If the Land switch now has no affect on the rhythm, but does when the pot is turned *down*, the pot is simply wired in reverse. Disconnect the wire going to the outside lug of the pot and reconnect it to the opposite outside lug.

Again with a rhythm playing, turn on a Land switch and slowly turn up the pot. At a certain point the rhythm will become disturbed, transformed into new musical entities. Fine-tune the pot for further strange variations on meter and composition. Try the same switching changes with no rhythm selected at all, and with only an instrument voice chosen as you play the keyboard. Do the same with the demo tune playing. More unusual output should result (keep trying variations in switching—some combinations will be more productive than others).

Next, try the same experiments with the Sea switches. Adjust the Catalyst dial. Play the keyboard by holding down a number of keys; notice the effect of various key combinations on the bent music. With the correct combinations of Land and Sea switches, and with the Catalyst dial tuned just right, many astounding musical patterns will result.

As always, if any control doesn't have the desired effect or does nothing, trace its wiring back to the board. Usually you'll be able to discover a bad solder joint or other trouble somewhere along the way.

Reassembly

By now you've closed the case and are ready to go busking. Replace the screws and that's it. Beware again of the trimmer masquerading as a Phillips-head screw on the back of the case. Resist the temptation and concentrate on the real screws only.

Musicality

After you become familiar with the somewhat tricky operation of the bent MT-140, you'll be able to coax all kinds of experimental music out of it. Live through a mixer (so that you can monitor things pre-house mix) or meticulously adjusted in the studio, many inspiring, interesting, and often surprising impromptu chance compositions will present themselves over time. Have a recording system handy!

Project 10: Casio SK-1 Aleatron

After early electronic sound synthesis had moved beyond the single-minded target of synthetic speech, next to come under examination was the synthesizing of musical instruments. An electronic oscillator's output would be adjusted to a certain waveform. This signal would be filtered (tone adjusted) and sent into an envelope generator to give the sound an attack time, a decay time, a sustain time, and a release time—theoretically modeling the audio content of an instrument voice. If things went well, you might get an approximation of the violin, piano, or flute you were after. And if things went *really* well, your colleagues in the Musician's Union would start sending hate mail—a sure sign of success.

So went the general school of thought until an unorthodox idea struck within the synthetic instrument community: Why not design an instrument around actual recordings of real instruments?

I have in my collection both the Mellotron and the Optigan, famous and, indeed, infamous examples of this technology. The nice-sounding Mellotron is a keyboard instrument whose magnetic tapes are actuated to run against a play head when the tape's key is pressed. Yes, there's a separate tape and play head for each key on the keyboard. King Crimson, the Moody Blues, Pink Floyd, YES, and many other strong groups used the thick, orchestral Mellotron sound.

The 1970 Optigan, another keyboard design, looks to instruments recorded on optical discs for its sound source. Though usually sounding a little queasy, the Optigan discs also held recordings of rhythm sections and bass lines, assuring that you'd keep all guests simply spellbound with entertainment. Tom Waits, when not misbehaving at the pub, is known to play the Optigan ("Frank's Wild Years"). The Optigan didn't slip through Devo's fingers, either, as heard on "Beautiful World" (Devo's EZ Listening Disc).

But what your guests really wanted was dogs barking "Jingle Bells." And somehow, on the other side of the world, Casio knew it. Digital sampling made this miracle of dogkind possible.

Casio's SK-1 musical keyboard brought the power of digital sampling to the public in the same big way that Texas Instruments introduced speech synthesis in its Speak & Spell. The SK-1 was real news, and was even heard on the air in groups whose members were never to admit its puny presence. But the SK-1 did show up everywhere, including, eventually, the bender's bench (see Figure 23-1). If you're a fan of deep and strange sound assemblages, this is your beast. I consider it one of bending's best targets.

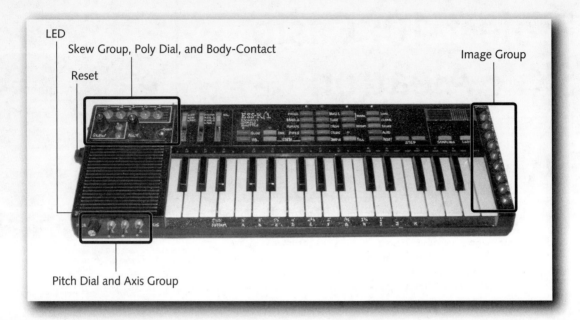

LED

Skew Group, Poly Dial, and Body-Contact

Image Group

Reset

Pitch Dial and Axis Group

FIGURE 23-1: The Casio SK-1 Aleatron

Many people consider the Casio SK-1 to be the Holy Grail of vintage samplers. I was very fond of my first and didn't bend it until I had a "junker" to experiment on first. The deeper you bend, the closer you come to fate (burnout *or* greatness). You might want to expose your cherished SK-1 to only the vibrato body-contact bend. Then again, you might want to go further than the guide here instructs! There's much more to bend within the SK-1.

Parts

Before you begin this project, be sure to have all these parts at hand.

- 1 Casio SK-1 Keyboard
- 18 miniature SPST (or SPDT) toggle switches
- 2 mini 500K potentiometers
- 1 10Kmini potentiometer (try to find one 15 mm in diameter; try vendors at end of master parts list in Chapter 13)
- 1 miniature N.O. (normally open) pushbutton switch
- 1 body-contact

Open It Up!

Remove all screws from the bottom and carefully separate the halves, keeping in mind that the battery compartment is located within the bottom half of the case. After marking where the wires go (including polarity), de-solder them from the battery compartment.

Circuit at First Glance

This circuit looks a little intimidating at first glance. What's worse, you need to connect loads of wires to the little black IC staring you in the face. And check out how tiny the IC's pins are!

Don't worry. Here's a great example of "follow that trace" (see Figure 23-2). Once again, you'll connect to this IC by following its traces to other spots where the traces open into larger targets for soldering.

Step-by-Step Bending

You've gathered all the needed parts and you've opened things up. You've studied the diagram. Now, step by step, you'll complete the transformation. Here's how.

Choosing a Cool Control Layout

In Figure 23-2, you'll see the controls laid out in several groups. Feel free to deviate from this plan—there's more room along the back of the case if you don't mind removing the circuit board to get to it. In fact, mounting the row of switches to the right of the keyboard might require you to remove the circuit board unless your soldering tip is long enough to reach the switch lugs.

Although the crashed SK-1 usually resets by your turning the power switch off and back on, if you find a need to remove a battery to reset your particular SK-1, consider locating a reset switch, as shown in Figure 23-1. See Appendix A.

Case Considerations

If you're mounting the controls as shown in Figure 23-1, no special considerations will be needed for the case. There is, however, the possible need to relocate a component on the board to make room for your new controls. This is covered under "Control Mounting," following in a moment.

Marking the Board

Looking closely at Figure 23-2, mark the circuit board at every point at which you'll be attaching a wire. Be careful when marking the rows of contacts on the left side of the circuit. Count in from the closest corner of these rows to locate the point you'll be soldering to (these rows are actually pins of ICs on the other side of the board). Note that the long wire from Axis switch three goes to the third pin down on its row.

Drilling Holes

Mark all holes with a pencil and follow up with drilling all pilot holes with a $\frac{1}{8}$" drill bit. Using your hand bore, bring holes up to size for the various components you'll be mounting.

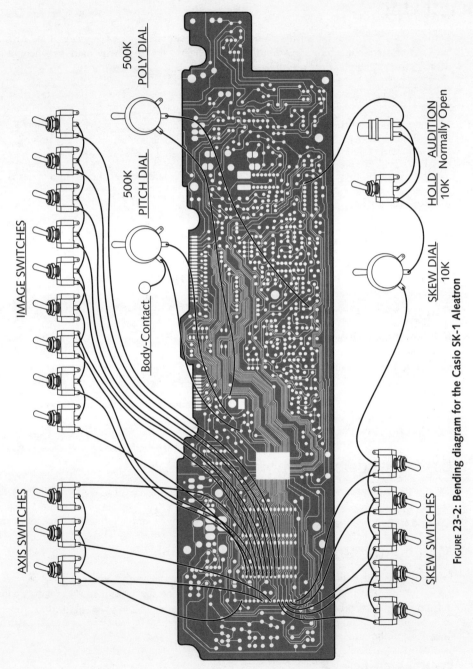

POLY DIAL
500K

PITCH DIAL
500K

Body-Contact

IMAGE SWITCHES

AXIS SWITCHES

HOLD   AUDITION   Normally Open
10K

SKEW DIAL
10K

SKEW SWITCHES

Figure 23-2: Bending diagram for the Casio SK-1 Aleatron

In some hole locations, you may find that using the hand bore will not allow completion of hole size. The circuit board may block deep enough penetration and the holes will still be too small. Either remove the circuit's retaining screws and keep it out of the way as you ream the holes larger, or carefully use a burr bit of the correct diameter for the final hole.

Painting

To paint the SK-1, you'll probably want to remove the keyboard and set it aside in a safe place. Just about all toy keyboards (and some professional ones, too) are molded in separate, black-and-white "trees." Each tree consists of a plastic rib to which the keys are molded. What you need to keep in mind here is that these trees are fragile. Breaking a key off the tree is easy to do unless the tree is handled and stored with care.

Also remove and set aside the three sliding switch handles and all the soft rubber pushbuttons, too. You might want to make a note of where the different-colored buttons were as well as all the titles printed on the case.

Control Mounting

If you're mounting any controls in the very convenient space located above the speaker, you might need to relocate a large electrolytic capacitor on the opposing circuit, depending on the one Casio used in your model. If so, note the polarity and relocate to another area where it will no longer interfere with control mounting (or the closing of the case!). This is best done by first removing the circuit board. Do so with care, and try not to flex the ribbon cables attached to the circuit board more than needed. See Appendix A for component relocation.

Space is tight inside the SK-1. Drill all holes with caution and care.

Soldering

Of all the projects in this book, this is the one for which *quick* and *precise* soldering is a must. Most of the soldering is done to heat-sensitive areas (IC pins), and the soldering points are very close together. Be sure to use thin-gauge wire (the 25–30 AWG wire-wrap wire is perfect for this). Be sure to use thin solder. Be sure that your soldering tip is super clean. And be sure to use as little solder as needed to make your connections—that is, unless you like the tedium of removing solder from between all the traces that *weren't* supposed to be soldered together. Take my word for it: You don't.

Testing

Testing the bent SK-1's response, if I were to try to cover everything the SK-1 now does, might fill a book the size of the one you're now reading! Rather than try to negotiate such a second edition with Wiley, I'm condensing the idea into the smallest tidbit possible: Experiment! Here are the basic steps.

Close the case halves and replace a few screws. Turn all the new toggle switches off. Turn the pots to their middle positions. Install batteries. Turn the SK-1 on using the main power switch.

If the SK-1 doesn't operate as usual, turn the new pitch pot and try again.

If you still receive no response, try holding the new reset button in while you turn the keyboard on. If this works, you've installed a "normally open" instead of a "normally closed" pushbutton. Install the N.C. switch instead.

If you still get no response from your revered SK-1, you've either wired things wrong (check for accuracy as well as solder's overflowing the target into other printed circuit traces), or you've heated things up too much and fried something inside an IC. The remedy for the former is a soldering fix. But for an unfortunate few, the remedy for the latter is many long years of psychotherapy.

So, let's hope your SK-1 is now working fine. Select the piano voice. Try touching the body-contact for a nice vibrato. Check your wiring if there's no response.

Flip the first Image switch and listen to the piano voice again. That's different, isn't it? Try all the switches one at a time to audition what they do. Some will work in combination with each other; some will not.

Now try another voice and go through the Image switches again. Some voices will work better than others. Some voices might even be silent.

Turn all the Image switches off. Now try the same experiments with the Skew group and finally the Axis group. The Skew group all run through a potentiometer, so be sure to try turning this pot as you audition effects.

The pitch dial, reset button, and envelope LED's operation should be checked. Evaluate wiring if any switch set does not operate or if any other control misbehaves. Remember to switch leads on the LED as a first troubleshooting step. And be aware that not all switch combinations will work together. Don't press your luck. Press reset instead.

Reassembly

In testing the bent SK-1, you've already nearly reassembled it. Replace the remaining screws and hope that you don't have to reopen it someday to prove to airline security that it's only art. By the way, airline security always asks me if what I'm carrying is dangerous. I look concerned and say, "Well, sure, it's art." This gives me a chance to practice my licks while I'm detained.

Musicality

Anyway, the bent SK-1 can produce everything from simple tonal changes to deep experimental passages. Personally, I enjoy the curtains of evolving musical noise that occur when several switch groups are used in combination. Samples as well as programmed voices can be mutated into these odd choirs of alien voices and instruments.

Highly recommended here is charting! In early, non-normalized synthesizers, in which all the modules were connected by patch cords, the complexity of the patches often grew nearly incomprehensible. Trying to recreate the patch without something to go on would be impossible. The "patch chart" was born. Usually little more than a drawing of the synthesizer's "face" (modules and patch bays), inputs and outputs were connected by pencil on the patch chart in replication of the true patching on the synth.

If you draw up a simplified representation of the bent SK-1 Aleatron, with all switches represented as boxes, you can check off and use circles to represent dials that you can draw pointers within, and you can then chart switch and dial positions to recreate an SK-1 Aleatron "patch."

Project 11: Mall Madness

H ere's a quickie. Mall Madness is a speaking game whose speech routines can be bent out of shape—*way* out of shape—with a couple of really easy bends. This project begins the final group of eight projects representative of the off-beat circuits abounding at the secondhand shops. These eight instruments begin simply enough, with a couple that require only a switch or two to function. They end with perhaps the most demanding project in this book, the Casio digital saxophone.

Mall Madness is usually found as a boxed game, although from time to time the plastic centerpiece of the game containing the circuitry will be found separately (see Figure 24-1). My guess is that somewhere a new doll design didn't work out and someone thought, why waste the plastic? The peach-colored but hard-lined circuit housing is very hard to miss.

Parts

Before you begin this project, be sure to have all these parts at hand.

- 1 Mall Madness game
- 2 miniature toggle switches
- Line-output array as desired (see Appendix A)

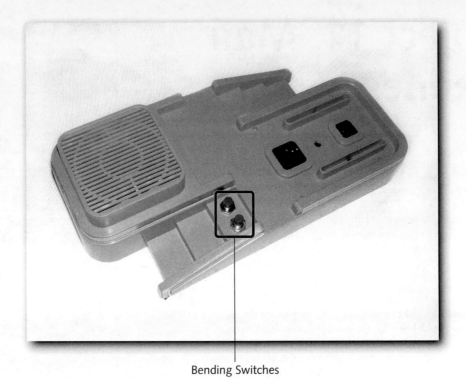

Bending Switches

FIGURE 24-1: The Mall Madness Machine with top buttons removed

Open It Up!

Removing a few screws from the bottom of the housing exposes the circuit. Be prepared for the fact that both the speaker and the circuit board will be loose.

Circuit at First Glance

This time, all connections can be made on the top of the board, now staring you in the face. You have the choice of soldering to an IC pin or the lead of a capacitor (see Figure 24-2) for one of the bends, because an IC trace connects them all on the reverse side of the board.

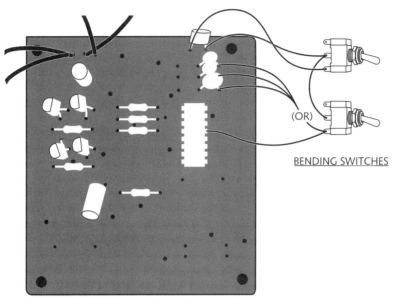

FIGURE **24-2:** Bending diagram for the Mall Madness Machine. Note that there
are three alternate points, all capacitor leads, that the bottom bending switch
can connect to in case you'd rather not solder to the IC pin. Any of the four
points works just the same.

Step-by-Step Bending

You've gathered all the needed parts and you've opened things up. You've studied the diagram.
Now, step by step, you'll complete the transformation. Here's how.

Choosing a Cool Control Layout

Considering the size of the housing, there's really not much accessible extra room inside. My
positioning of the two mini pushbutton switches on the particular "step" location (see Figure
24-1) takes best advantage of the space available.

Case Considerations

All you're really dealing with here are component and mounting tool clearances. There are no
ribs to remove. The plastic is easily workable and, if you're mounting your switches as shown,
everything should be a snap.

Because the line output will be derived from the speaker terminals, it makes sense to locate the output jack nearby. Before deciding upon a control layout, look at your line-out options in Appendix A. You'll need to add an output jack of some type, and you have the options of installing a speaker cut-off switch and even an envelope LED (if the speaker signal will drive it; it may not be strong enough). If the line-output is too strong, you'll want to include a trim pot in the scheme, also depicted in the drawing in Appendix A.

Marking the Board

In the bending diagram (refer to Figure 24-2), you'll see that you have several options as to where to solder one of your switch leads. As mentioned previously, it can go to either an IC pin or one of three capacitor leads. Any of these four locations will work because they're all connected on the other side of the board. The capacitors will be less heat-sensitive than the IC, so if the capacitor leads are large enough to solder to, mark one of these.

Drilling Holes

Mark your two hole locations, allowing component and tool clearances. Mounting two hex-nut-mount components too close to each other will interfere with tool usage: Your crescent or socket wrench will hit the adjacent component's hex nut if located too close.

In the illustration, you see two mini toggles. Pushbutton switches could be used as well if you prefer to not keep the switches in an "on" state.

Using your ⅛" bit, drill your pilot holes and enlarge them with your hand bore, as usual.

Painting

Stripping-down the housing (as seen in Figure 24-1) is easy and the first step in painting. Removing the circuit is simple—just de-solder the battery wires (first mark the polarity of the wires so that you'll know how to reconnect them). If you want to paint with the circuit still inside the housing, that's fine. Just be sure to mask both sides of the circuit first to keep paint from interfering with areas that you'll need to solder to as well as the printed circuit traces that serve as switch contacts on the bottom (trace side) of the board.

Control Mounting

Mount your two switches in the position you decided upon. If you've located them as shown, you shouldn't encounter any real problems.

Soldering

If you've chosen to solder to the IC pin instead of one of the capacitors, make this connection in the "quick and precise" style—you don't ever want to overheat or use too much solder when soldering to IC pins.

You'll notice that you're wiring another "1 to X" bend. One wire is common to both switches and attaches to a spot on the board that supplies two bends, each handled by a separate switch.

Line-output circuitry is the same in all projects if not otherwise noted. See the line-output scheme in Appendix A and assume that the speaker in the diagram is the speaker in the project.

Testing

Start the game and begin pressing your switches as the speech routines are running. The speech should be disrupted, and in its place you'll hear the kind of things that would keep me, at least, in the mall longer: strange alien languages instead of cell phone sales hype.

Reassembly

There is the chance that the original game's pushbuttons will separate from the board if the board falls away from its mounting posts inside the case. If this happens, carefully insert the buttons within their plastic frames and position the circuit board over them, locating the board upon its four pegs. This will align things properly; replace the bottom of the case and drive in the screws.

Musicality

I always like streams of connected yet disconnected phonemes (or in this case *allophones*, electronic rather than acoustic building blocks of speech). I find these musical in themselves, especially when time-shifted downward. Perhaps one of the resistors on the board sets the pitch. You might try some resistor substitution here, as described in the Cool Keys project (Project 15) and depicted in Appendix A.

Project 12: The Inverter

It's been a while since the target circuit of this project was marketed. So long ago that I forget the original name. I call it an Inverter because the bending "inverts" (actually convolutes) the original sounds onboard. In reality, the bending seems to trigger all kinds of new noises, really extreme noises, in crisp high fidelity.

You'll be looking for what you see in the photo (see Figure 25-1) but minus the round metal grill. There's a chance the toy will still have its elastic straps intact—a dead giveaway. The straps allowed the wearer to fasten the device to a wrist so that, while playing baseball or riding a bike, appropriate sound effects could be triggered via an internal motion sensor.

When the toy was in bike mode, you heard motors revving up and such. In baseball mode, you were treated to the crack of a home run, the crowd cheering, tobacco spitting, and other big league noises.

Inside the case was a small "tilt" switch, similar to the classic shut-down switch on pinball machines. Here, though, when the tongue of the switch is jolted and its motion closes the connection, it plays one of the built-in sounds. In keeping with the motion-sensor aspect of the design, I install a tiny mercury switch across the trigger points for the bent results (which is why you see no controls on the outside of the case—there aren't any). If you'd prefer a small pushbutton switch to initiate the bent responses, that's fine. Just wire the switch to the same points that the mercury switch goes to.

Figure 25-1: The Inverter (with internal mercury switch)

Parts

- 1 Inverter toy
- 1 miniature N.O. (normally open) pushbutton switch

or

- 1 mercury switch (minis available online)
- Line-output array as desired (see Appendix A)

Open It Up!

Two screws hold the case together. One is obvious—right in the middle of the back half of the case. The other (note that it's of a different type) is within the battery compartment. Keep these screws separate and memorized. Mistakenly screwing the wrong one into the middle of the back case will strip the threads that the correct screw needs to locate against.

Circuit at First Glance

You're looking at the back of the board (see Figure 25-2). Note that in the diagram the top bending wire connects to one of a set of four contacts arranged in a tight row. This is a row of pins coming from an IC on the other side of the board. Look closely and be sure to solder your wire to the second pin over from the left.

Note that the thumbwheel at the bottom left is the volume control. Just for curiosity's sake, try to figure out the potentiometer on the circuit.

See the three central terminals? These are the three soldering lugs common on all potentiometers. Look closely at the two extra soldering lugs as you turn the dial all the way on and off. See what's happening? The outer lugs attach to a switch mechanism that, although mounted to the pot, has nothing to do with the resistance levels of the pot at all. This is the built-in power switch and is a very common potentiometer style.

Note

If you wanted to bypass such a pot in a bending project (don't be surprised) you'd need to do two things. First, you'd need to break the three central connections going to the pot, just beyond the rivets (a tiny burr bit on your Dremel drill accomplishes this). Your new pot would solder to the original connections. Second, leaving the original thumbwheel turned all the way down (which turns the thumbwheel switch off), you'd wire a toggle switch to the switch's original two connections. Now you've extended both functions of the pot to your remote components that do the same thing exactly, but now in the form of a full-size pot and power switch capable of being mounted on a custom case of your choice.

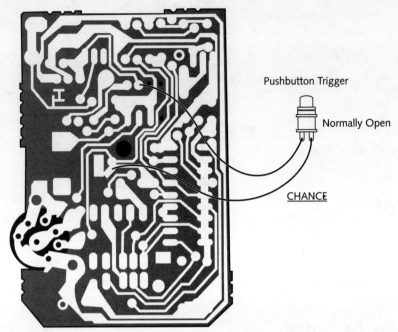

FIGURE 25-2: Bending diagram for the Inverter. Illustration depicts a mini pushbutton switch capable of being mounted on the case. Optional is to use an internal mercury switch instead (as in Figure 25-1), allowing actuation by simply tilting the instrument to one side.

Step-by-Step Bending

You've gathered all the needed parts and you've opened things up. You've studied the diagram. Now, step by step, you'll complete the transformation. Here's how.

Choosing a Cool Control Layout

If you'll be installing a mini mercury switch inside the case, try to orient its closed (ON) position to trigger in a direction opposite the direction of the mechanical motion sensing switch already built into the toy. A process of trial and error is the key to discovering this relationship.

If you intend to mount a miniature pushbutton switch to trigger the chance responses, be sure that the switch clears the top of the circuit board and is in a good spot for your fingertip to naturally fall upon.

Because the line output will be derived from the speaker terminals, it makes sense to locate the output jack nearby. Before deciding upon a control layout, look at your line-out options in Appendix A. You'll need to add an output jack of some type, and you have the options of installing a speaker cut-off switch and even an envelope LED (if the speaker signal will drive it; it may not be strong enough). If the line-output is too strong, you'll want to include a trim pot in the scheme, also depicted in the drawing in Appendix A.

Case Considerations

As long as you're careful to mount your miniature switch (mercury or pushbutton) in a place that does not interfere with either the board or the closing of the case (look before you leap), all should be well. The plastic used in these devices is a little more brittle than most you'll run into. Keep all tools under good control if you intend to alter the case at all.

Marking the Board

As noted previously, study the circuit closely and be sure to locate the correct row of IC pins for the top wire connection. This is the second over from the left (see Figure 25-2). Mark the terminal with your Sharpie fine-tipped felt marker. Look below to the bottom connection point and locate the printed circuit trace shape, as shown in the figure. Mark this point (or trace, because you can solder anywhere along it).

Drilling Holes

No holes need to be drilled if you're using an internal mercury switch. If you prefer a mini pushbutton switch, first use your pencil to mark the location of the intended switch (again, verify all clearances within the case first). Next drill your pilot hole with a $1/8$" bit.

Because the plastic is a little bit more fragile than most, drill with care. Use a medium-high speed and go slowly. Exert only as much pressure as is needed to make slow progress. When the hole is drilled through, carefully and slowly enlarge to the correct size with your hand bore. As before, go slowly and exert minimal pressure against the plastic.

Painting

After the circuit is removed from the plastic case, the case becomes easy to paint. Because the battery compartment door already has very tight tolerances, it becomes mandatory to use the thinnest of paint coats so as to not cause the door to bind.

Control Mounting

If you're going the internal mercury switch route, just be certain that the switch's leads don't come into contact with other circuit areas when you close the case. It's never a bad idea to enclose such an added circuit board component within a sandwich of paper (or tape) to keep it insulated.

If you're mounting a small pushbutton switch on the case, you'll probably be tightening a hex nut. Grab the crescent wrench of correct size.

Soldering

"Quick and precise" is a soldering description that pays off continually in bending, and here's its place. Your top connection is to an IC pin, where it's always wise to think "quick and precise." Be sure that the point of your soldering tip is super clean any time you're soldering to IC pins.

The bottom connection is much less critical. Pick your spot along the indicated trace and solder your wire. These leads go to the switch of your choice.

Line-output circuitry is the same in all projects if not otherwise noted. See the line-output scheme in Appendix A and assume that the speaker in the diagram is the speaker in the project.

Testing

Testing is quite simple. Install the 9-volt battery and actuate a sound by moving the instrument swiftly sideways. When a sound is initiated, either tilt the instrument to trigger your mercury switch or press your pushbutton. If you've wired things correctly, you'll hear an amazing assortment of bizarre sonic mutations when your new switch closes. Keep repeating the procedure. I still, many years later, hear new things from Inverters.

If there's no response to the new switch, check your wiring. When bent correctly, this circuit never crashes. It just keeps on cranking out more strange sounds.

Reassembly

Because you haven't added much to this circuit, reassembly is straightforward. Just be sure that you're not pinching any wires (or your internal mercury switch) when you bring the case halves together. As covered previously, don't mix up the two screws that hold the case closed. The thinner one goes into the middle of the instrument's back.

Musicality

Sometimes the difference between music and sound effect is slight. Although I'd lean quickly toward sound effect here, some of the bent voices are quite musical. If sampled and then worked with in the electronic studio, the various voices of an Inverter could, with ease, find their way into compositions.

Project 13:
Electronic Rap Pad

In a classic black-and-white science fiction movie from the 1950s, *Earth Versus the Flying Saucers*, an alien message is revealed as the batteries in a portable tape player die. What had seemed to be spaceship noise resolved into spoken word when slowed down, as the motors of the tape deck drew to a stop. Mayhem, of course, resulted from the message's having been initially missed, and moviegoers got to see what they came for: the destruction of Washington, D.C.

Benders know that slowing things down can reveal alien voices. But the principle for us usually works in reverse. Rather than taking the unidentifiable and slowing it down to make it recognizable, we slow down the recognizable and make it unidentifiable. Often this process is as simple as replacing a resistor with a potentiometer, exactly what we're about to do to the Electronic Rap Pad (see Figure 26-1).

Parts

Before you begin this project, be sure to have all these parts at hand.

- 1 Electronic Rap Pad
- 1 5M (or 10M) pot
- 2 body-contacts
- Line-output array as desired (see Appendix A)

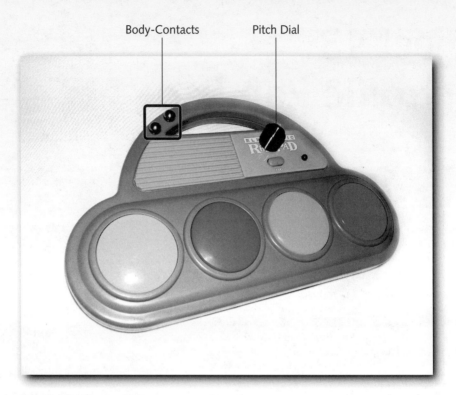

Body-Contacts   Pitch Dial

FIGURE 26-1: Circuit-bent Electronic Rap Pad

Open It Up!

Nothing could be easier. Remove the small, Phillips-head screws from the back of the case and the two halves will separate, remaining connected only by the wires coming from the battery compartment.

Mark on the circuit board where the battery compartment wires connect. (Be sure to note which color went where or you could reverse the polarity of the power supply when you reconnect it later.) De-solder these wires so that you'll be able to work on the front half of the instrument more easily.

Observe that there's a strip of circuit-board–like material spanning the area beneath the four "pads" of the instrument. If you're not careful, this strip will casually fall out of place and allow all four pads to clatter to the floor, making one of the instrument's most interesting sounds so far. Still, you might want to set the pads aside, being certain that their central contact inserts remain in place.

Circuit at First Glance

Not much to see! Two resistors, one electrolytic capacitor, and one transistor are the only obvious components (see Figure 26-2). Next to the transistor, on the other side of the board, is the IC, an epoxy dot type rather than the rectangular "DIP" (dual in-line pin) IC more often seen.

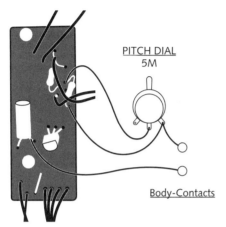

FIGURE **26-2: Bending diagram for the Electronic Rap Pad. The 5M potentiometer replaces the resistor to its left (see Appendix A for a drawing of resistor replacement).**

The little bit of work needed to complete this project is all done on the side of the circuit board facing you. Because you've already removed the battery compartment wires, you'll have complete access to the board. Wiring things should be a snap.

Step-by-Step Bending

You've gathered all the needed parts and you've opened things up. You've studied the diagram. Now, step by step, you'll complete the transformation. Here's how.

Choosing a Cool Control Layout

There are two spots for a potentiometer. If you have a small pot, you can mount it as shown in Figure 26-1, right above the Demo switch. If you're using a full-sized pot, check out the area between the center two pads and toward the control panel. You will need to use a couple spacers between the pot and the panel on the inside to keep the shoulders of the pot from hitting the nearby edges (use washers or an extra hex nut for this setback).

Aside from the pot, all you'll need to mount are a couple of body-contacts and an output jack (the latter only if you want to feed the signal to an amplifier; see Appendix A for the generic line-out circuit). The body-contacts will easily locate anywhere along the handle or at its base, as will the line output as long as you're using either an RCA jack or a ⅛" mini jack.

Because the line output will be derived from the speaker terminals, it makes sense to locate the output jack nearby. Before deciding upon a control layout, look at your line-out options in Appendix A. You'll need to add an output jack of some type, and you have the options of installing a speaker cut-off switch and even an envelope LED (if the speaker signal will drive it; it may not be strong enough). If the line-output is too strong, you'll want to include a trim pot in the scheme, also depicted in the drawing in Appendix A.

Case Considerations

The Electronic Rap Pad is enclosed in a very simple case. The plastic is easily workable and presents no problems to the bender. There are no ribs to remove and no unusual clearance situations. In fact, it's probably the simplest case design in this book, and the easiest to work with.

Marking the Board

This project is so simple that you might not even need to mark the board! After all, you're just replacing a resistor with a potentiometer. One body-contact goes to one of the pot connections; the other goes to the "leg" of a capacitor (okay, may as well go ahead and mark that leg).

Drilling Holes

After checking for component and mounting tool clearances, mark hole positions with a sharp pencil. Use the ⅛" drill bit to drill your pilot holes. Finally, use your hand bore to enlarge holes to component size and finish them off with your de-burrer.

Painting

By now it will have become obvious how to disassemble the instrument. The front and back halves, as well as the four pads, can be painted as you like, and all go back together without a problem at all. See Chapter 12 for details on painting.

Control Mounting

Mount all controls. See "Choosing a Cool Control Layout," previously in this chapter, for special potentiometer considerations. Mount the potentiometer so that its lugs will be available for easy soldering.

Soldering

It's not necessary to remove the resistor from the board to replace it. Just clip one lead from the board, leaving enough lead near the board to still be able to solder to. See the bending diagram (refer to Figure 26-2) to target the correct resistor. Also see Appendix A for the generic resistor replacement circuit.

Line-output circuitry is the same in all projects if not otherwise noted. See the line-output scheme in Appendix A and assume that the speaker in the diagram is the speaker in the project.

Testing

After all the bent connections have been made, re-solder the battery compartment wires to their circuit locations. Load your batteries.

There is no on/off switch—just press a pad against the internal contact strip and the pad's sounds should start. Turn the new potentiometer to tune the pitch of the pad samples.

While a sample is playing, try bridging the body-contacts with a finger. The pitch should change (this will be less obvious on slowed-down samples).

If either the pot or body-contacts fail to function, trace your wiring to find the problem. If the pot speeds things up when you turn it down, it's wired in reverse. Simply rewire it by de-soldering the wire going to its outside lug and re-solder it to the other outside lug.

Reassembly

With the contact strip in the correct position behind the four pads, align the front and back halves of the instrument. Check to be sure that none of your new wiring is in the way. Close the halves and replace the small screws.

Musicality

Slowed-down digital voices sound best with a little reverb and a bit of tone polishing using an equalizer. As with a didgeridoo, an aboriginal wind instrument, the bent Rap Pad is now a drone machine, and drones are a constant in world music. How you approach the drone in your own music is entirely up to you. You might want to use drones as backgrounds to lay other sounds on top of, or you might want to monumentalize drones as hypnotic foreground elements. Experiment!

Project 14: Bent Book Strip

Y ou wouldn't think that something so simple could be so cool! If you visit any charity store, such as Salvation Army or Goodwill, you're bound to find the heart of an exceptional experimental sound-field generator hiding behind cartoon faces over in the kid's book section. Where, exactly? Right inside those sound strips that edge any number of electronic kid's books, full of cartoon noises and other digital sound effects waiting to be bent (see Figure 27-1).

The book strip bend is one of the simplest in the project section. All you're doing is substituting a variable resistor for a fixed resistor on the circuit board, similar to the switch-out in the previous project. Nonetheless, as shown in Figure 27-2, this simple design can be worked up into a complex and, in fact, spellbinding instrument. More on the Dworkian register can be found in Chapter 11.

It is only within a narrow range of their playback speed that the samples within these strips—character voices, vehicle sounds, goofy noises—are recognizable. Above and below the normal speed, these sounds take on new meanings. Sped-up sounds are now insect chatter and alien radio static. Slowed-down sounds become animal growls, earthquakes, and thunderstorms. Imagine the sound fields possible with four of these bent strips, sent through an equalizer, through reverb, and out to a surround set of four powered speakers (again, see the Dworkian Register shown in Figure 27-2 for an instrument designed to do this).

Pitch Dial   Line Output

LED   Speaker Switch

Body-Contacts

FIGURE 27-1: Book strip removed from book and now triggered by a steel ball. A textured surface has been placed over the original cartoon-like drawings to keep the ball from rolling away from the sound-activating area it's been positioned over.

The example strip I discuss here is from a book titled *The Noisiest Day*. There are, however, many different book strips available to this process. In every case, all you need to do is replace the correct resistor (there will be only a few) with a large value pot (5 to 10M). If you're working on a strip that doesn't match the one pictured (as is likely), just clip one lead of the target resistor and test a pot across the electrical connection that the resistor made. This is a very common and very productive bend, possible in many circuits.

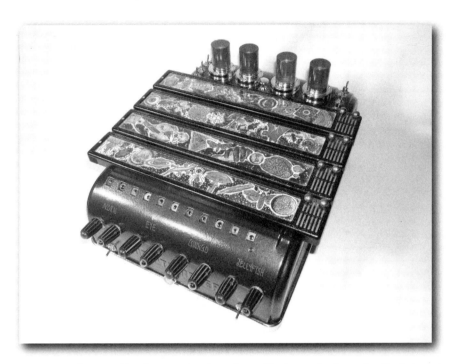

FIGURE 27-2: The Dworkian Register is a four-channel instrument whose circuitry is based upon the configuration of four independent book strips.

Parts

Before you begin this project, be sure to have all these parts at hand.

- 1 book strip
- 1 1M pot (miniature, unless you're mounting controls in a larger add-on case)
- 2 body-contacts
- Line-output array as desired (see Appendix A)

Open It Up!

Here's a job for one of your tiniest Phillips screwdrivers. Set all the screws aside in a safe place. Be sure to remove the screw holding the battery compartment drawer closed; the case will not open otherwise.

Circuit at First Glance

The circuit at first glance, in this case, is the bottom of the circuit. Your work needs to be done on the other side of the board—you need to remove a resistor—so grab the tiny Phillips screwdriver and remove the screws holding the board in place.

Note that the membrane keypad contact ribbon will be freed when the board is removed. Be sure to observe how the board/ribbon connection is made, and align things correctly during reassembly. It's really very simple.

Now you're looking at the top of the board (see Figure 27-3), and the resistor you need to remove is accessible. If your board does not match the one pictured, you need to experiment a little to get the right one.

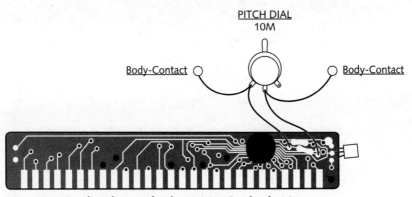

FIGURE 27-3: Bending diagram for the *Noisiest Day* book strip

Step-by-Step Bending

You've gathered all the needed parts and you've opened things up. You've studied the diagram. Now, step by step, you'll complete the transformation. Here's how.

Choosing a Cool Control Layout

All you need to mount is a potentiometer and possibly a body-contact or two. The only open area is next to the speaker, and that area is very limited. Unless you have a very small potentiometer, you might be looking at mounting the strip on another housing and using that housing to accommodate the new controls (again, see Figure 27-2). On the other hand, if you frequent the surplus warehouses as often as I do, you might be able to dig up an ultra-mini 10M pot and mount it next to the speaker, along with your pitch-bend body-contacts (refer to Figure 27-1).

Because the line-output will be derived from the speaker terminals, it makes sense to locate the output jack nearby. Before deciding upon a control layout, look at your line-out options in Appendix A. You'll need to add an output jack of some type, and you have the options of installing a speaker cut-off switch and even an envelope LED (if the speaker signal will drive it; it may not be strong enough). If the line-output is too strong, you'll want to include a trim pot in the scheme, also depicted in the drawing in Appendix A.

Case Considerations

If you're unable to find a pot small enough to fit inside the case of your book strip, consider attaching the strip to another case of some kind. If you do have a small pot, locate it next to the speaker after you decide where the body-contacts, output jack, and LED will be located (if any of these additions will be included in the design).

Marking the Board

As in the Electronic Rap Pad project (Chapter 26), you need to replace a resistor with a potentiometer. If you're working with a strip other than the one pictured, you'll need to find the right resistor to target. Rather than mark the board, first you'll need to locate the right resistor to replace.

Simply clip a resistor's lead in the middle, halfway between the resistor's body and the board. Bend the resistor a little bit away from the board to be sure that the resistor is no longer in contact with the snipped lead (see Appendix A for a picture of resistor replacement).

Now, using two small alligator clip wires, clip a potentiometer (try a 5M) between the two points that connected the resistor to the board (refer to Figure 27-3). Turn the pot's shaft both ways and see whether the pitch of a triggered voice changes. You may have to extend wires from the cut resistor to the pot and reassemble the circuit/ribbon junction each time you test until you find the right resistor to replace.

With this simple design, it may not be necessary to mark the board. If you do search for LED or body-contact points, it's not a bad idea to go ahead and mark them. This always keeps things as clear as possible. And when you begin running wires here and there, even simple circuits can become confusing.

Drilling Holes

Using your pencil, keeping in mind control and tool clearances, mark your spots for the pot, the RCA output jack, and any other controls you intend to mount.

Use your ⅛" bit to drill pilot holes. Be extra careful this time. The plastic in these sound strips is often very thin and weak as well. Bring all holes up to size with your hand bore; finish hole edges with your de-burrer.

Painting

After all holes are drilled, it's an easy job to paint the frame that surrounds the membrane key-pad. Keep the battery door in place while you paint only as long as your paint coat is to be thin—a thick coat will paint the door shut. It's always possible to paint the battery door separately. If you choose to do so, be sure to mask all but the outside exposed area of the drawer with tape. Any build-up of paint on the inner door surfaces will interfere with the smooth operation of the drawer.

Control Mounting

Although you're working in a tight space, there are no real concerns here. Just be sure to choose locations that will allow you as much room for tool and component clearances as possible. If you're mounting the controls within an add-on housing of some kind, think about ergonomics: How will it be most comfortable to play the new instrument? Place and mount the controls as your custom layout allows.

Soldering

As long as you're using the thin wire-wrap wire that I've had you working with all along, there are no special considerations here. After you've found the correct resistor to replace (it is a hit-or-miss search), be sure that you're soldering your potentiometer wires to the correct circuit spots. If you've found LED and body-contact points, follow the circuit traces and solder to areas the least likely to invite trouble if there's a solder overflow.

Line-output circuitry is the same in all projects if not otherwise noted. See the line-output scheme in Appendix A and assume that the speaker in the diagram is the speaker in the project.

Testing

With the membrane strip attached and active again, go through all the voices to be sure that each still works. If some have gone silent, this is probably a ribbon problem. Be sure that the ribbon connecting the membrane keypad to the circuit is aligned correctly.

All voices should now be tunable from super-high to way down low (the higher the pot value the lower the voices will go). Your LED should strobe with the sounds. The body-contacts should bend the pitch when touched. The line output should supply a nice, clean signal to your audio system. Check your wiring if anything's not working as it should.

Reassembly

If you're working with an added case to allow more room, simply reassemble the membrane and frame and attach the whole thing to your larger housing. Run your wires to the controls within the added case and run all tests as outlined previously. If you're using the tiny case only (as in Figure 27-1), be very careful to keep wires away from the moving drawer of the battery compartment, and inside the case as you bring the front and back sides of the case together. Replace the small screws and retest all functions.

Note Consider this project a likely candidate for battery supply replacement. In the Dworkian Register (refer to Figure 27-2), I've replaced the tiny button batteries with three D cells for driving the circuits and the bank of high-brightness LEDs. See the ending of Chapter 9 for battery-swapping instructions, but build the new compartment right into the instrument's case.

Musicality

My instant feel for this type of sound system falls toward the ambiance generator (again, see the description of the Dworkian Register in Chapter 11). The unusual voices of the bent sound strips are great as mysterious punctuations to other music, and, using body-contact vibrato, can even become alien lead vocalists if you're hanging out with the wrong crowd.

Project 15: Cool Keys Bent Music Stick

W ho needs a mechanical piano keyboard to play the good ol' equal-tempered scale? No one, or so thought Diversified Specialists Inc. To stand by this conviction, around 1993 DSI introduced "Cool Keys," a stick-like guitar-shaped instrument capable of playing mellow bell-like tones over a span of two octaves by means of pressing a long, flat membrane colorfully marked with the notes of the scale (see Figure 28-1).

Instead of the usual abandonment of the sharps and flats, often omitted in little kids' music makers, they're included. The only things that keep this little instrument from doing bigger things are its limited frequency range, no vibrato bar (pitch bender as on a real electronic guitar), and its lack of an output jack. Why don't we remedy all this and a little more?

Parts

Before you begin this project, be sure to have all these parts at hand.

- 1 Cool Keys
- 2 miniature SPST (or SPDT) toggle switches
- 1 5M (or 10M) potentiometer
- 1 body-contact
- 1 red LED
- 1 1µF electrolytic capacitor
- 1 10µF electrolytic capacitor
- Line-output array as desired (see Appendix A)

Pitch Dial

LED, Bending Switches, and Body-Contact

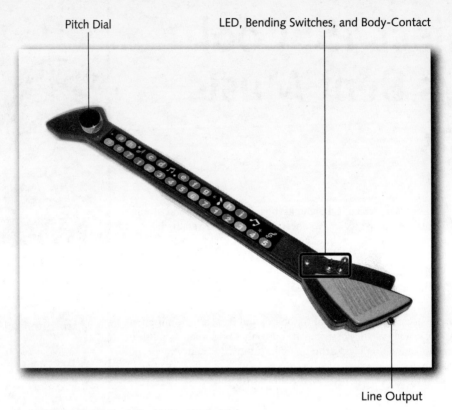

Line Output

FIGURE 28-1: Circuit-Bent Cool Keys Music Stick

Open It Up!

Remove the eight Phillips-head screws on the back of the instrument. Carefully separate the instrument halves. Be extra cautious here because the battery compartment resides on the back half of the case and is connected to the circuit by the frailest of wires.

As soon as you have the back off, it's a good idea to de-solder the two wires from the battery compartment. Be sure to mark on the battery compartment which wire went where (red is positive; black is negative). Now, working on the circuit will be easier and carry no risk of your (or your cat's) ripping the battery wires off the circuit. Nonetheless, remember or make a note of where the battery wires connected to the circuit in case a wire comes loose. As on the battery compartment, note polarity as well.

Circuit at First Glance

This is a small circuit but it holds no particular problems in bending (see Figure 28-2). Space is generally tight around the circuit and speaker area. Work cautiously and use small components.

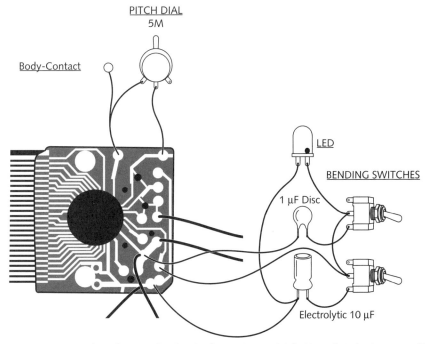

FIGURE 28-2: Bending diagram for the Cool Keys Music Stick. Note that the two capacitors can be soldered to the switch lugs. In this application their polarity is noncritical.

Step-by-Step Bending

You've gathered all the needed parts and you've opened things up. You've studied the diagram. Now, step by step, you'll complete the transformation. Here's how.

Choosing a Cool Control Layout

You'll put your switches on the lower body of the case, along with an LED. Exactly where you place the switches is up to you. You'll be able to use a full-size pot if you mount it in the "tuning head" above the neck. My usual design goes something like what you see in Figure 28-1.

Case Considerations

If you do mount a full-size pot in the head of the instrument, you might, before reassembly, need to trim away a plastic rib from the back of the case before the two halves will meet. Don't worry about the ribs up and down the neck—they won't interfere with the two wires coming from the pot.

Because the line output will be derived from the speaker terminals, it makes sense to locate the output jack nearby. Before deciding upon a control layout, look at your line-out options in Appendix A. You'll need to add an output jack of some type, and you have the options of installing a speaker cut-off switch and even an envelope LED (if the speaker signal will drive it; it may not be strong enough). If the line output is too strong, you'll want to include a trim pot in the scheme, also depicted in the drawing in Appendix A.

Marking the Board

Using your fine-tip Sharpie marker, make a tiny dot next to each circuit area you'll be soldering to.

Drilling Holes

All holes start with your pilot bit of ⅛" or so in diameter. Switch and body-contact holes go on the top face of instrument just as shown. Note that the RCA output jack goes on the bottom half of the case.

Painting

Masking the membrane keypad is probably a good idea if you want to keep track of the notes you're playing, though this never seems to bother most of the musicians I jam with. If you need to remove the yellow plastic grill, put a burr bit in your Dremel drill and grind away the yellow dots inside the case. The grill should then pop off.

Control Mounting

If you're using the control array depicted (refer to Figure 28-1), there should be no real concerns. The LED is, in my example, simply mounted in a snug hole.

Try to mount the pot far enough back on the ribbed section of the tuning head so that its nut tightens down on the grooved section of the plastic rather than at its edge.

Soldering

For the potentiometer to function correctly, it's necessary to remove the small resistor from the back of the board, exactly opposite where the new pot's leads connect. I was able to heat one of the resistor's leads from the back of the board with my soldering tip while prying it away from the board with a very tiny screwdriver. Only one of the resistor's leads need be removed from the circuit. See Appendix A for resistor replacement.

If the resistor is too close to the board for this, unscrew the retaining screws and lift the board from its mount. Be extremely careful not to stress the ribbon cable coming from the circuit. With this better access, you should be able to heat the resistor's leads from the back of the board while you carefully pry it from the board with a small pliers.

As you've probably guessed, the original resistor set the pitch of the instrument. Because you're replacing the resistor with a variable resistor (the potentiometer), you'll now be able to tune the instrument over a wide range. So far, in fact, that the instrument will now play notes from above your hearing range down to deep, earthy growls.

Be sure to solder your wires to the LED as quickly as possible. Remember that LEDs are heat sensitive and are candidates for heat-sink usage.

Line-output circuitry is the same in all projects if not otherwise noted. See the line-output scheme in Appendix A and assume that the speaker in the diagram is the speaker in the project.

Testing

Testing is very straightforward. Unlike the crash-prone Incantor series (Projects 1 through 5), the bent Cool Keys is rock-steady. Just put the batteries in and play away, flipping all switches and touching the body-contact as you're so inclined.

The pitch dial should tune the instrument from super deep to beyond hearing range. If it's not working right, be sure that you've removed the resistor on the other side of the board (see the "Soldering" section, just prior).

If a switch doesn't work, check the wiring. Same with the body-contact. If it doesn't change the instrument's pitch when touched, check to be sure that it's soldered to the correct point. In this case it can go to either the indicated point on the board or to the pot lug connected to the indicated point on the board; each provides the same connection.

As usual with LEDs, if it's not lighting, the first thing to try is reversing the leads. If this doesn't work, try an LED of a different style (lower power requirement or higher efficiency/brightness).

Reassembly

Line up the two instrument halves, being careful not to trap the two wires coming from the newly installed RCA jack. Replace the screws.

Musicality

What you now have is a nice little carry-around synth stick. The frequency range is expanded. You now have cool note sustains, body-contact vibrato, and a line output for your grunge pedals. Plug it in and see what it now can do. Cool by itself, it also shines as a basic wide-frequency controller for an effect chain.

Project 16: Fisher Price Electronic Womb

Somehow, I like the idea of womb sounds, bent or not. The womb is already an experimental sound chamber, or at least it would be considered such if we were able to return there now and listen to the world outside from within.

What Fisher Price was trying to do with this already odd-sounding circuit was to re-create womb sounds and produce a comforting sonic environment for newborns. The less-than-comforting-looking device was to be hung on the side of baby's crib, turned up, and left to soothe baby while Mom and Dad tried to get some rest. Cover your ears, little one: It shall soothe no more (see Figure 29-1).

Parts

Before you begin this project, be sure to have all these parts at hand.

- 1 Fisher Price Electronic Womb
- 4 miniature SPST (or SPDT) toggle switches
- 1 1M potentiometer
- 1 100K potentiometer
- 1 red LED
- Line-out array as desired (see Appendix A)

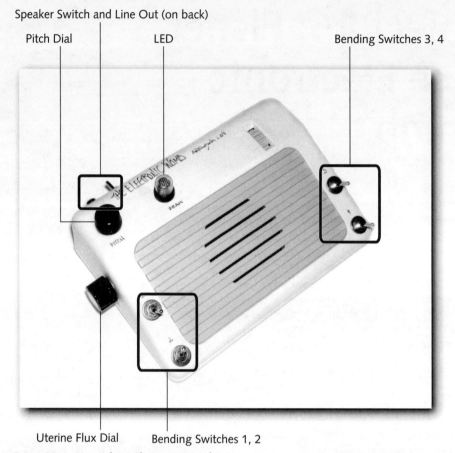

Speaker Switch and Line Out (on back)

Pitch Dial   LED   Bending Switches 3, 4

Uterine Flux Dial   Bending Switches 1, 2

FIGURE 29-1: Circuit-bent Electronic Womb

Open It Up!

Removing the six deep-set screws from the bottom half of the case facilitates opening. The wires from the battery compartment keep you from moving the case bottom very far away from the top half of the case. You should be able to slip the 9-volt battery clip through the slit in the battery compartment and thereby allow case halves to be separated entirely.

Circuit at First Glance

You're looking, of course, at the back of the circuit board. Everything you need to get to is right there. Better yet, all connection points are good sized and will permit soldering with ease (see Figure 29-2).

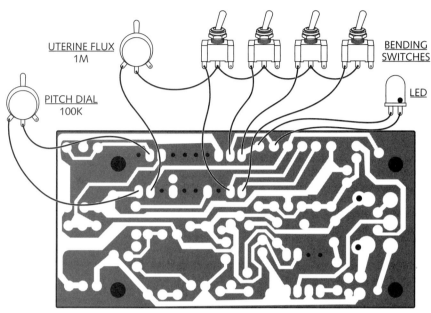

FIGURE 29-2: Bending diagram for the Electronic Womb

Step-by-Step Bending

You've gathered all the needed parts and you've opened things up. You've studied the diagram. Now, step by step, you'll complete the transformation. Here's how.

Choosing a Cool Control Layout

Because this is an instrument more "set" than played, ergonomics will not hold the same role as in keyboard instruments, whose keys and controls must meet the hands in the best and most comfortable way possible. I've laid out the controls in a somewhat symmetrical fashion, but there is room enough to vary this arrangement.

Because the line output will be derived from the speaker terminals, it makes sense to locate the output jack nearby. Before deciding upon a control layout, look at your line-out options in Appendix A. You'll need to add an output jack of some type, and you have the options of installing a speaker cut-off switch and even an envelope LED (if the speaker signal will drive it; it may not be strong enough). If the line output is too strong, you'll want to include a trim pot in the scheme, also depicted in the drawing in Appendix A.

Case Considerations

If you bend this circuit beyond the instrument shown, you might need room for more controls. Although I've been able to avoid mounting anything on the gray plastic speaker grill, there is room to do just that. If you do choose to mount controls on the speaker grill, be sure to locate the mounting holes in the center of a ridge, not a valley. This will allow best placement of the panel hardware (washer or hex nut) for tightening down against the grill.

Marking the Board

Using your fine-tipped Sharpie pen, mark the circuit board in accordance with the wiring diagram in Figure 29-2, leaving a dot everywhere you'll be soldering a wire.

Drilling Holes

After marking all control positions (remember to be sure of component and tool clearances), drill your pilot holes with a ⅛" bit. If you'll be mounting your LED within a pilot lamp housing whose collar or hardware will hide your mounting hole, the hole itself is not so critical. If you intend to mount your LED all by itself, snugly fitted into tight holes, use the bit that makes the exact-size hole for the LED, and, as usual, be sure to drill the hole from the inside of the case toward the outside.

After your pilot holes are drilled, open them to final size with your hand bore (or a burr bit if the hand bore won't penetrate far enough). Finish all hole edges with your de-burrer.

Painting

When all holes are drilled and the nameplate is removed, painting is fairly easy.

Note   Nameplates and sticker-based graphics are easily removed by first warming them with a heat gun (small forced-air heater available in hobby and craft shops), peeling them off, and finally removing any adhesive left behind with either rubbing alcohol or, much better, Goo Gone, a commercial glue solvent.

You'll also want to remove the circuit board to get the thumbwheel out of the way first.

As outlined in Chapter 12, paint the case in thin, even coats. One advantage to painting the Womb Synthesizer is its initial color: white. This allows good color fidelity, the white being a perfect "undercoat" to any color you want to choose (a dark case needs a white primer coat to make top colors look right; again, see Chapter 12 if you need a refresher).

Control Mounting

Mount your controls using the correct tools. As long as you've given yourself ample room, you should encounter no special problems because there's good working space within the instrument.

If you're mounting your LED without placing it into a housing and the LED is still a little loose in its hole, that's okay. Heat the hot-melt glue gun and put a dab of hot glue on the back of the mounted LED. Hold the LED in place as the glue sets while watching it from the front side, keeping it central in the hole.

Soldering

All soldering points are large enough to work with without much trouble. You will be soldering to several points that are actually IC pins (look closely at the two rows of pins at the circuit's upper-left quadrant). The connections you'll make should be of the "quick and precise" style, as usual when soldering to IC pins.

Line-output circuitry is the same in all projects if not otherwise noted. See the line-output scheme in Appendix A and assume that the speaker in the diagram is the speaker in the project.

Testing

Attach a battery to the clip, turn the thumbwheel up, and listen to the usual womb sounds with all your new controls turned off.

Next, play with the pitch dial, the uterine flux dial, and the four bending switches (uterine flux works only in conjunction with the four bending switches). The LED should flash with the loudest sounds.

If any controls don't have the desired effect, trace their wiring and connect correctly. As always, look for solder mistakes that might be connecting components or traces not meant to be connected to each other.

Reassembly

Slip the battery clip back into the battery compartment. Being sure that all your added wires are running inside the instrument in such a way as to escape being pinched, align the instrument halves and bring them together. Replace all screws.

Musicality

As it begins, so it continues. Originally a drone machine, it's still a drone machine. But now as you turn the Uterine Flux dial and change the bending switches, you'll be able to transform the organic in-utero symphony into variations upon a theme. Similar to playing a digeredoo or a bowl gong (whose edge is rubbed to bring forth a tone), you'll be experimenting with extending intrinsic hypnotic tonalities. I often use such drones as constants to "set" musical scenes, to introduce other, more animated sounds within.

Project 17: The Harmonic Window

I suppose I *was* acting suspiciously. After all, I was forcing my hand up the behinds of fluffy toy animals, one after the other. All the while, I was being carefully watched by a security guard and several alarmed parents, peeking at me between rows of discontinued Barbie accessories. Weird situation, no? The opening, right beneath the tail, was tight, made for a child's hand. But with my hand inserted up to the wrist, I was able to feel the lump I was after. Remember: Art is always a fine excuse for antisocial behavior.

I continued through the row of these fuzzy animals, checking each one. By the end of the row, I'd filled my cart with eight or so creatures, the smiling face of each one now looking a bit in shock. What clued me in to this line of toys was a bit of palm reading: If the palms of these animals read "PLAY" and "RECORD," they were then fair game for my hunt. That is, as long as they still contained the sampler that my frisking was all about (see Figure 30-1).

I no longer recall exactly what the creatures looked like. There were maybe three versions: a dog, a bear, and perhaps a cat. But each had the same tattoo on the palms: PLAY and RECORD. And this is what to look for amidst all the abandoned toy animals at the secondhand shops. You *will* be able to see into the future with a little such palm reading, with the future in this case being a super-cool bent sampler (see the Harmonic Window on the book cover and in Figure 30-2; read more about it in Chapter 11).

Picture Switches   Line Out   Record, Play, and Stack buttons   Pattern Switches

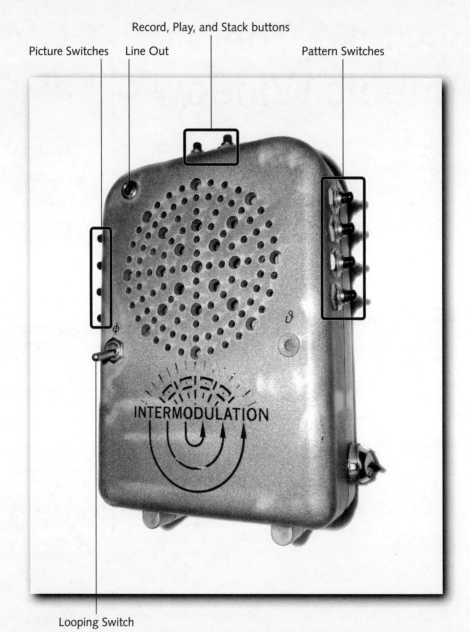

INTERMODULATION

Looping Switch

FIGURE 30-1: Miniature Harmonic Window made from the battery compartment/circuit of a sampling teddy bear

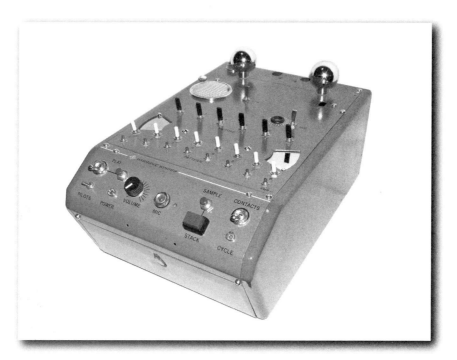

FIGURE 30-2: The Harmonic Window

What's so interesting about this instrument is that bending the circuit results in a sampler that can then stack samples in a row, something it could not do before. In addition to this, the stack of samples can then also be transformed in various mutations of themselves, resulting in percussive loops and phrases of extraordinary content. The transformed sample stack is often accompanied by odd swept frequencies and noise anomalies creating surprising complexity out of the simple sample source of, for example, a pencil tap and a couple shouted syllables.

Parts

Before you begin this project, be sure to have all these parts at hand.

- 1 Harmonic Window sampling circuit
- 9 sub-miniature N.O. (normally open) pushbutton switches
- 1 short wire (clipped resistor lead) for the stacking jumper

FIGURE 30-3: Ultra-simple Sample Stacker—the same circuit housed inside an emptied Polaroid camera body

Open It Up!

Reaching up into the animal allows you to grasp the sampling unit and remove it from the cavity, though it's much more fun with a security guard watching. Remove the batteries. Remove the six screws securing the back of the case and carefully separate the two halves.

If your circuit has a cardboard shield protecting it, remove the shield and replace the two small Phillips-head screws (don't overtighten; the plastic here is soft and strips easily).

Circuit at First Glance

You'll see at the top of the board where the wires that went to the paw switches connect. Your new Play and Record switches will connect, of course, to these same points. Study them and note which contacts go to which switches.

Look closely at Figure 30-4. The most important bend on the board is the small jumper visible just below the central IC. Adding this jumper makes the stacking function possible.

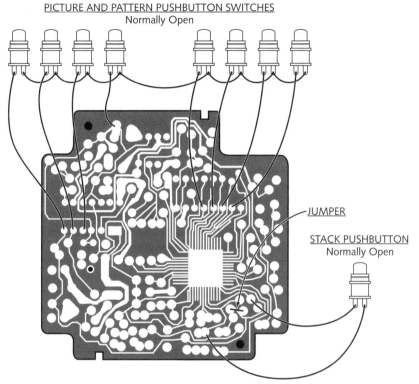

PICTURE AND PATTERN PUSHBUTTON SWITCHES
Normally Open

JUMPER

STACK PUSHBUTTON
Normally Open

FIGURE 30-4: Bending diagram for the Harmonic Window. Wire your new Record and Play pushbuttons to the points where the originals connected (not shown in illustration; just follow the original wires and solder to the same points).

All your connections will be made on the back of the board; there are no unusually critical soldering situations to deal with.

Step-by-Step Bending

You've gathered all the needed parts and you've opened things up. You've studied the diagram. Now, step by step, you'll complete the transformation. Here's how.

Choosing a Cool Control Layout

Your control layout will depend upon whether you want to create a miniature carry-around version (refer to Figure 30-1) or instead mount everything within a new case of some kind. The case I used for the large version on this book's cover and shown in Figure 30-2 is from an "AnsaPhone," an early telephone answering machine that recorded on reel-to-reel tape.

If you're going the same route, that of a larger case, you'll need to examine your new case and note where the circuit will be mounted, where your new battery supply should go, and where all the controls will be mounted.

If you'll be attempting to mount all controls on the original case (as shown in Figure 30-1), you'll have tight clearances, but you'll have space for sub-miniature components all along the top and both sides.

Controls are divided into two groups, described next.

The Record and Play Group

This group consists of four switches. Record and Play are the original controls offered by the sampler. The Stack switch allows live samples to be stacked end to end. The Loop switch simply keeps the Play switch turned on. The three of these that are pushbuttons (Play, Record, and Stack) I keep near each other, as seen at the top of the miniature instrument in Figure 30-1.

The Picture and Pattern Group

The Picture and Pattern buttons are the Stack modulators. Mounting these as opposing pushbuttons on the miniature model makes sense because they'll need to be pressed in combination with each other to play the instrument. With them mounted thus, the instrument can be played with one hand.

If you're mounting the Picture and Pattern groups on a larger case, consider using both pushbutton and toggle switches wired in parallel (see Figure 30-2) for one or both groups (for parallel wiring of toggle and pushbutton switches, see the section in Appendix A called "Audition/Hold Array"). The toggles will allow you to lock in the combinations you're playing.

You also need to consider the built-in microphone. This is a powered "electret" microphone and needs to be unblocked, available to pick-up sounds. You have two choices. First, you can just mount the entire sampler housing inside your larger case with the sampler's microphone situated right behind a hole in the larger case. Second, you can remove the microphone from the original housing and extend it away from the circuit board, to be mounted all by itself somewhere on the new case.

If you choose to do this, de-solder the wires going to the microphone and extend them as needed with longer wires, using shielded cable.

The larger case allowed me to use big body-contacts. If you're using a metal case, be sure to mount the body-contacts through rubber grommets to keep them insulated from the case. If you don't, they'll short-circuit to each other, won't work, and might put the entire circuit at risk. (Body-contacts are not marked in Figure 30-4, but you'll find several kinds if you search.)

Because the line output will be derived from the speaker terminals, it makes sense to locate the output jack nearby. Before deciding upon a control layout, look at your line-out options in Appendix A. You'll need to add an output jack of some type, and you have the options of installing a speaker cut-off switch and even an envelope LED (if the speaker signal will drive it; it may not be strong enough). If the line output is too strong, you'll want to include a trim pot in the scheme, also depicted in the drawing in Appendix A.

Case Considerations

If you're using a custom housing of some kind, you'll just be laying things out as the case might suggest. Consider the idea of the groupings mentioned previously in "Choosing a Cool Control Layout."

If you'll be mounting all the controls on the original housing, you'll need to remove the circuit board first. After the board is removed, you'll have access to the sides and will be able to work with the case more easily.

With a little planning, you can use the hole that the original Play and Record wires ran through to mount one of your switches.

Marking the Board

Using your Sharpie marker, mark the board's connection areas as shown in Figure 30-4. Be sure to locate the correct areas to mark—there are a few rather confusing circuit areas where similar traces run next to each other.

Drilling Holes

Use a pencil to mark all hole locations after checking to be sure of component clearances as well as mounting tool clearances. Using your $1/8$" pilot drill, drill all holes. Further open these holes using your hand reamer. If ever the reamer won't penetrate far enough to achieve the desired hole diameter, carefully enlarge the hole with the side of the spinning bit (as described in the section "Hole-Drilling Fixes" in Chapter 10), or use a burr bit of the correct diameter.

Painting

If you're sticking with the original plastic case, painting is a cinch. Don't overpaint or the battery compartment door might stick. Keep the halves separate, as well as the battery door, and paint lightly.

Control Mounting

All 12 switches on the mini version are simple to mount using small crescent wrenches. If you're working with a larger housing and bigger controls, the standard techniques of control mounting will probably do. Unusual mounting situations, however, require a little thought. Don't rush the process and you'll probably work out unique control interfaces that will take things further than you might have thought.

Soldering

As mentioned previously, the circuit is pretty well packed with traces running close to each other. Be careful to use only as much solder as needed to create a good electrical connection. Excess solder on a cramped circuit always invites trouble.

Think similarly as to soldering the switch lugs. Any excess solder will further encroach upon the circuit after it's re-mounted.

You'll need to remove the four wires emerging from the circuit's top right—the wires that went to the original Play and Record switches. Heat these junctions as you gently pull on the wire. The wires will come off and the junctions will be ready to accept your new wires.

Use a short piece of bare wire (such as a small section of a lead clipped from a resistor) for the jumper (see Figure 30-4). Solder one end to one side of its connection. Let it cool and then bend it down against the remaining connection point. Clip it to exact length (no overhang) and quickly solder it to the second point, completing its placement.

Note that the wire that connects all the Picture and Pattern switches to each other does not connect to the circuit anywhere, as it would if it were the common wire in the "1 to X" bend (see Chapter 9). Here, you're connecting switches to each other in many combinations rather than all to a single circuit point as in the "1 to X" bend.

You might want to study both circuits until you understand how they work. If you already understand the "1 to X" bend, think of this bend, accordingly, as the "X to X" bend. Each of these two bends will be frequent discoveries within bending. See these two circuits side-by-side in Appendix A, along with further explanation.

Line-output circuitry is the same in all projects if not otherwise noted. See the line-output scheme in Appendix A and assume that the speaker in the diagram is the speaker in the project.

Testing

I'm using the mini model we've been discussing, but if you're working with another housing, the testing procedure is the same. Follow along.

Press the Record button to sample a sound. Press the Play button to listen to the sample. Now flip the Play toggle switch (looping) to lock the sample in a "repeat play" mode.

Each time the sample finishes, press the Stack button to record another sample at the first sample's ending. If you keep the samples short, you'll get a string of four or five samples running in a loop. If, however, you try to record beyond the time limit of the sampler's short memory, the stack will be erased and you'll be left with only the straw that broke the camel's back playing over and over again. Now that's adding insult to injury.

When you have a nice stack playing, try pressing different combinations of the Picture and Pattern buttons. Try pressing Picture and Picture, Pattern and Pattern, and Picture and Pattern switches together. The result should be unusual modifications of the stack in play.

Body-contacts, if you included them, should produce either a distortion or vibrato effect on the sound, depending upon the Picture/Pattern relationship.

Reassembly

Reassembly, in the example of the mini version, is a matter of being sure that the wires you've added are run away from the circuit and won't be pinched when the halves of the case are fastened back together. Likewise with any larger case you might be working with: Be sure that all wire runs are thought out well and non-interfering.

Musicality

In a sense, the Harmonic Window sampler is a one-man alien band. With the line output feeding a strong sound system, the wide frequency range and unusual musical sounds resulting from bending the sample stack take on a ferocious space of hard-driven, sharp-edged musicality. Heavy metal, grunge rock, electronic minimalism, and music concrete intertwine within the Harmonic Window's bent compositions. If unusual meters or near-insane, intense experimental music is of interest, this may well be your dream come true.

Project 18: Hyper Sax

O nce again, Casio took the lead in bringing cool new music technology to the masses with the mid-1980s release of the Casio DH 100 (see Figure 31-1). Breath control was the feature that caught everyone's attention in this six-voiced, sax-shaped digital horn synthesizer. It gained many fans, including The Residents (who've used it in live performance).

For a while I thought that my bending might have destroyed my DH 100. It began to squeal, with the real voices heard only faintly in the background. Turns out this is a failure common to the DH 100. If you perform the following bends on your DH 100 and it begins to squeal at a later date, after you say "cool!" and then fear that that's all it's ever going to do, remember that this digital disruption was bound to happen.

Don't worry—I have a fix for this squealing. I'll outline the simple procedure in a moment (as well as a cool trick to get a different range of frequencies out of the DH 100 by crystal swapping).

Note You can purchase squealers on eBay and elsewhere for nearly nothing, with the seller assuming that the horn is beyond repair. If squealing is the only problem, worry not. You just found your bending target. Perform the fix, described later, and proceed to bend.

And now the warnings. This project involves bends that alter electronic balances that the circuit depends upon to correctly execute general instrument envelopes as well as breath control envelopes. Worse, the actual breath sensor assembly is prone to malfunction if disturbed, which is hard to avoid while you're working inside the case.

This project also involves precise circuit-board surgery (if you decide to go ahead and replace the envelope trim pot with a surface-mount pot) and, because of space restrictions, relies upon miniature potentiometers that can be somewhat difficult to obtain (using standard-size pots is possible with some careful alignment; read on).

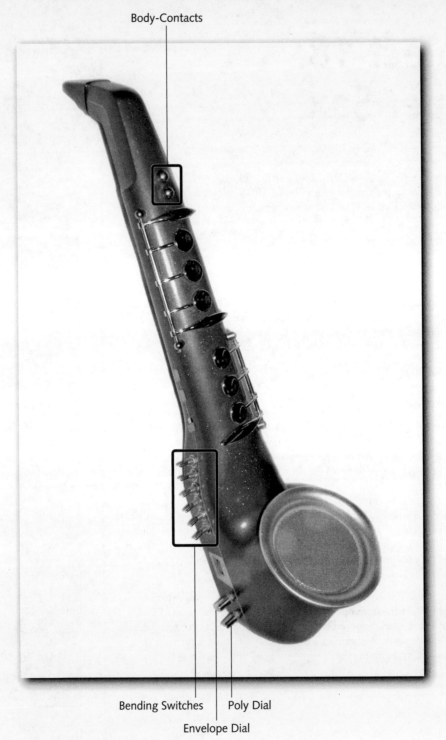

Body-Contacts

Bending Switches   Poly Dial

Envelope Dial

FIGURE 31-1: Circuit-Bent Casio DH 100

Still not challenging enough? Okay. How about the fact that this project additionally requires the most precise soldering of any in this book? Wires need to be soldered to SMDs (surface-mount devices: super-tiny resistors and capacitors) as well as to an ultra-slim IC trace. Still, everything's possible as long as you keep your soldering hand steady.

If you can dig up the small pots, and if your composure is not shaken by the thought of cutting printed circuit traces or soldering to diminutive circuit areas (or if you never play your DH 100 again, anyway!), then what's to lose? Be brave and tackle this project. The bends *will* extend the instrument wonderfully. And your band members will eventually thank you no matter how much they liked the squeal.

Parts

Before you begin this project, be sure to have all these parts at hand.

For the Main Bends:

- 1 Casio DH 100 Digital Sax
- 6 miniature SPST (or SPDT) toggle switches
- 2 miniature 1M pots (or full size; see text)
- 2 body-contacts

For the Capacitor Replacement Repair:

- 1 (or 2, as needed) 47 µF electrolytic capacitors (RS part #272-1027)

For the Crystal Expansion:

- 6 crystals of choice ranging from 1.8432 MHz (RS part number 900-5089) to 20 MHz (RS part number 900-5126)
- 1 six-position rotary switch (RS part # 275-1386)

Open It Up!

Opening the DH 100 is fairly simple as long as you keep things in place. Be aware of the fact that the breath control's sensor is very, well, sensitive. Treat it gently. Pull off the black mouthpiece, remove the screws, and proceed with caution. Don't let the screw in the battery compartment snooze. Get it out of there.

Be prepared for the small buttons along the side of the instrument to fall from their mounts and dangle around as you try to work. Observe how they were mounted and make a mental note.

Circuit at First Glance

As shown in the bending diagram (see Figure 31-2), all the bends are available on the top side of the board. However, you'll need to remove the circuit from its mounting posts to access the reverse side for cutting the traces coming from a trim pot (covered in a moment) and to perform the squeal repair, if needed.

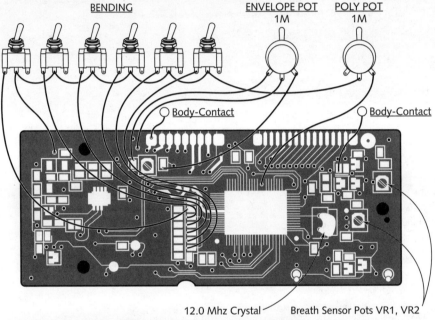

FIGURE 31-2: Bending diagram for the Casio DH 100. Experiment with other potentiometer values for different effects.

Notice the miniature SMD resistors (black, two lead), capacitors (terra cotta brown), transistors (black, three lead), and others such tiny components on the board. Each has an identifying title associated with it. Resistors are marked nearby R1, R2, and so on; capacitors are C1, C2, and so on; and transistors are coded T1, T2, and so on.

Look closely at an SMD resistor or capacitor. The shape is that of a tiny rectangular box, each end terminated by a minuscule metal cap. Hold your breath and turn on your internal gyroscopes: It is to these silvery caps you'll be soldering. The diagram shows SMDs as white blocks, but the wiring indicates to which side of the SMD (silver cap) you'll solder to.

Step-by-Step Bending

You've gathered all the needed parts and you've opened things up. You've studied the diagram. Now, step by step, you'll complete the transformation. Here's how.

Choosing a Cool Control Layout

If you've managed to find small pots, you'll be able to configure controls as shown in Figure 31-1. If you're using standard-size pots, read on. I cover that in a moment.

The two body-contacts, mounted as shown, require the removal of the contact strip inside the case opposite the instrument's keys. If you want to save the trouble of removing the strip, it's possible to mount the contacts elsewhere. Study the case and you'll see room here and there. Think about ergonomics and choose a good spot.

Case Considerations

It's what's in the case, not the case itself, that presents challenges. There's a lot in the way. The main circuit board itself blocks pot and switch mounting. As mentioned, the key contact strip blocks the pictured body-contact mounting. Luckily, both board and strip are easily removed. I get to that in a moment.

Marking the Board

Study the board and compare it to the wiring diagram. There are two soldering points that are not directly associated with components. These are through points—points where solder ports from the other side of the board come through to the top. One is to the left of the SMD capacitor "C1" at the board's top left (this goes to the poly pot as well as to the six toggle switches). The other such point is above and to the left of the trimmer marked "VR3" (variable resistor 3). This point is the furthest left of six such ports situated in a row. Counting inward from the upper right corner, IC pin 8 is the one you're after.

Using your fine-tip Sharpie marker, mark these points and the rest of the bending points as shown in the wiring diagram.

Note

Casio has abandoned the standard nomenclature of component codes on this board. Instead of "R," variable resistors are marked "VR," and instead of "Q," transistors are marked "T." Who knows?

Drilling Holes

Jump down to "Control Mounting" and read up on board removal and potentiometer choices. After you've decided what kind of pots you'll be using, you can decide where they'll be mounted and, accordingly, mark mounting positions with your sharp-tipped pencil. Look again at "Choosing a Cool Control Layout" for body-contact ideas.

After all positions have been marked (and any internal components removed, if needed—probably the main circuit board; be sure that all new switch and pot clearances are double-checked), all pilot holes can be drilled with your $\frac{1}{8}$" bit. Use your hand bore to enlarge holes up to component size; finish hole edges with your de-burrer. Remember that, with careful use, you can also use your Dremel and the correct-sized burr bit to enlarge holes. The burr bit will want to stray around the pilot hole. Be very careful to keep it on track.

Painting

A lot of disassembly is required to paint the DH 100. As with other complex instruments I've discussed, luckily this is simply procedural. All boards remove quite easily (screws or slots) and the actual key hardware is held in place with a minimum of screws. Keep all screws sorted as to origin, and make notes of any internal configurations you think you might need to refer to upon reassembly. When you have the unadorned case halves drilled out and cleaned and (probably) the speaker grille and its chrome rim masked, you're ready to paint. See Chapter 12 for painting tips.

Control Mounting

If you can't find the small pots needed to complete the design as shown in Figure 31-1, you may be able to use standard-size pots if you mount them in the bottom of the horn's bell, beneath the speaker. You'll need to be aware of the flexible drain tube that runs through this area as well as a mounting post that houses a screw to close the case at the bottom of the instrument. Be sure to meet clearance needs before you drill holes in this area.

To install the panel-mounted envelope pot, you need to cut three traces on the board's bottom (see "Soldering," next). After traces are cut and your new components are mounted, you can finally relocate the board and replace its mounting screws.

If your switch or pot lugs will be blocked by the board, be sure to solder wires to these lugs before remounting the board. Make these wires longer than you think you'll need them; they can be snipped to correct length as you later solder them in place.

Soldering

If you want to be able to reshape your voice envelopes, you need to replace a trimmer on the board (VR3) with a panel mount potentiometer (the envelope pot in Figure 31-2). Study the traces on the back of the board that connect to VR3. Using a sharp knife blade or very tiny burr bit on your Dremel, cut the circuit traces as close to the pot's leads as possible. Essentially, your aim is to isolate the trimmer from the circuit, because the new envelope pot solders to traces that once connected to the trimmer, completely replacing it.

 Note   It is not necessary to replace VR3, the envelope trim pot, with a panel-mount pot as shown in Figure 31-2. Although this bend allows cool variation in the DH 100's sounds, it does require precise circuit surgery, and no other bend depends upon the replacement of VR3 to operate. It's entirely up to you whether to include this touchy bend in the Hyper Sax bending scheme.

After all the new bending components have been mounted behind the board (pots and switches), remount the board.

Soldering now proceeds as indicated in the bending diagram (refer to Figure 31-2). It is imperative that you use a tiny soldering tip when soldering to the metallic ends of SMD components. And it's mandatory that the tip of your soldering iron be as clean as possible.

Quick and precise, our bent soldering standard, is needed here if anywhere. SMDs are a little daunting to solder to. But as said, all's possible here. Proceed with a steady hand, take your time, and use good lighting and proper tools.

Reassembly

The DH 100 needs to be reassembled prior to testing because it's just too cumbersome to test otherwise.

Be sure that all new wires along with both small side circuits and the drain tube are positioned correctly. The drain tube lower end and white rubber breath tube need to be located in their respective slots. Slowly bring case halves together. When both sides are nested to each other properly, replace the screws in the back half of the case and tighten. Finally, replace the black mouthpiece.

Testing

With both new pots centered in their range and all new switches turned off, insert batteries and turn the instrument on. Play the DH 100 as usual, but experiment with the breath control against the pot settings. Slowly introduce your bending toggle switches and observe their effect upon the circuit. Touch the body-contacts and listen to the pitch and tone changes they induce.

As always, check your wiring if anything seems wrong, or if the circuit ceases to function. Other than the squeal and often temperamental breath control, the DH 100 is rather hardy. If anything's not working properly, you can probably trace it to a bad solder connection and remedy things quickly.

Musicality

In contrast to many projects in this book, in which the bent instrument response is totally alien, the bent DH 100 still retains a strong degree of recognizable musicality. Although the bending switches introduce interesting pulses and stutters to the voice, and the dials bring forth varying overtones, envelopes, and textures, you're still playing the equal-tempered scale. You've simply created a breath-controlled instrument that is no longer strapped to horn sounds. Instead, you're now playing a breath-controlled, who-knows-what synth from Mars, and that's all the fun.

Special Section: Casio DH 100 Squeal Repair Session and Crystal-Swapping

Prior to the existence of transistorized circuitry, it was common to see vacuum tube testers in hardware stores, pharmacies, and electronics shops. Every home had replaceable tubes within, and tube shopping was a way of life.

Tubes, as do light bulbs, age and eventually fail. The first symptom of tube failure in audio circuits was often an audible buzz or hum. The suspect tube was flicked with a fingernail. If the hum responded, the tube was verbally demeaned, removed from the socket, and escorted to the drug store. With luck, the tube would test to be bad and the shopkeeper would have a replacement on hand (maybe even stored inside the back of the tester). Those were the days of home-fix circuits. Rejoice—those days are not gone.

Today, a likely culprit behind audio hum, buzz, or squeal is the capacitor. Especially "canned" capacitors, such as electrolytics.

The Squeal Fix

On the back side of the main circuit board, you'll find capacitor C39 near the far edge (see Figure 31-3). If your DH 100 is squealing, this is probably the culprit. It might even be leaking a fluid. You need to remove C39 from the board, clean the board, and replace C39 with a new capacitor. (C30 is also suspect if your DH 100 contains a C30—not all boards do; replace it too if replacing C39 does not solve the problem.)

Note   There's more than one version of the DH 100 circuit. It will be up to you to determine the actual location and capacitor type used for C39 or C30 if your board differs from the one shown in Figure 31-3. It will also be up to you to determine exactly where the new capacitors will solder into the circuit, because you might have options beyond the actual holes the original caps were mounted in and the ones shown in Figure 31-3.

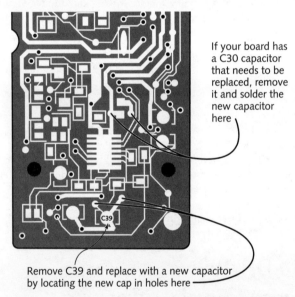

If your board has a C30 capacitor that needs to be replaced, remove it and solder the new capacitor here

Remove C39 and replace with a new capacitor by locating the new cap in holes here

FIGURE 31-3: The two suspect capacitors behind the common DH 100 squeal are C39 (shown) and C30 (absent from the board illustrated). After the offending capacitors are removed, the new caps can be soldered into the locations (holes) shown.

If your C39 has leads you can snip, snip them. Solder an electrolytic 47 μF cap to the same spots the original cap was attached to, or the alternate spots shown in Figure 31-3 (the original was a 33 μF, but the 47 μF will work fine and is usually easier to find—Radio Shack part number 272-1027). It's fine to use a 33 μF if you happen to have one. Either way, be sure that its voltage rating is above 6 volts (voltage rating usually appears on electrolytic caps) and that you observe the correct polarity when installing. (There will be a "+" sign within a circle somewhat near the positive mounting hole for the original cap; the new cap will be marked as to its own polarity with a "+"or "-" stripe pointing to one of its leads.)

If there's no room to snip the leads of C39, you'll have to heat the cap's leads from the other side of the board while you pry the component away from the board. Heat one lead at a time and pry a little bit at a time until the cap is free from the board. Work very carefully and check when you're finished to be sure that no damage was done to any printed traces and that no solder "bridges" (shorts caused by melted solder) remain between traces. Clean the board with a little alcohol after you're done. As before, replace the old C39 with your new cap, being sure to note polarity (see Figure 31-3 for soldering locations).

Crystal Swapping

Now for the Crystal swap. On the top side of the board, you'll find "CSA 120 MT 18" situated to the right of the main IC in its robin's egg–blue case (and visible in Figure 31-2). This is a 12.0 MHz ceramic crystal needed to set the DH 100's frequency (learn to recognize crystals in circuitry for the following generic experiment).

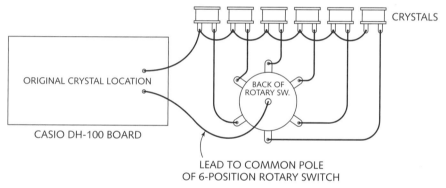

FIGURE 31-4: Use a multiposition rotary switch to select crystals and therefore expand frequency coverage for your DH 100 Hyper Sax.

On my human voice synthesizers, I use a rotary switch to choose among crystals so that the voices can speak from bass to tenor, and then some (see Chapter 11). If you're into the idea of extending the frequency range of your DH 100, see Figure 31-4 for a super-simple crystal switcher.

The crystals can be soldered to the rotary switch poles and chosen with the turn of a dial. All you need to do is to clip (or de-solder) the original crystal from the DH 100 board and replace it with the switch/crystal circuit as shown. Experiment with crystals both above and below the original 12.0 MHz. "Canned" crystals are okay to use instead of the ceramic original; such are available through Radio Shack in frequencies from 1.8432 MHz (part number 900-5089) to 20 MHz (part number 900-5126). These once-common items are no longer shelved at Radio Shack and will probably need to be special ordered or found elsewhere. Try to incorporate a series of crystals that will provide you with a nice frequency range to work with, lowest to highest.

Tweaking the Breath Control

You might also want to experiment with the trimmer pot labeled VR2 in case the breath sensor ever seems to malfunction (see Figure 31-2). Remember, though: If you've replaced the envelope pot as in the bending instructions, unless you have the new envelope pot set to exactly where its original trimmer was set (unlikely), the circuit will already be out of balance, and changing the positions of VR1 or VR2 might just send you deeper into the unknown . . . maybe not a bad thing. But beware.

The idea, given that you're working on a horn operating within a reasonable proximity to norm, is to "null" the circuit to a point just prior to its sounding *without* blowing. Hold a note down and turn the trimmer VR2 until the held note can just be heard. Now back the trimmer off until the note just disappears. This should set the breath response to the most sensitive and correct position. VR1 also comes into play here, but after the balance between VR1 and VR2 is lost, things can get ugly fast. If you do want to experiment with these pots, mark their initial positions first to give you some kind of reference point to try to return to later when the horn refuses to make a sound at all!

In this last project, you've grappled with some of the most demanding situations a bender will face: mechanical components to keep straight, circuit board surgery, micro-soldering, tight clearances, and touchy sensor technology. All in all, a fun project and one that will prepare you for just about anything you'll find as you bend into the future.

Closing Words

Here we are at the end of this discovery journal, and at the end of our brief travels together. But in truth, it's just the beginning of your adventure. Any guide to circuit-bending can but scratch the surface of this intrinsically vast art. There are literally thousands of new instruments waiting to be discovered. Circuit-bending is about as open an art as any to exist.

To gain a little perspective and at the same time realize the importance of working in experimental, difficult art, let's place you, the bender, within history.

Divergent art requires divergent thought, an attribute easily overlooked and often underestimated today. However, bent art, thought-provoking radical art, was not always so easily produced. The work you're doing *historically* and *sociologically,* as well as artistically, is important.

We trace prevailing western social order to the Middle Ages, to the period of the Renaissance in particular. During the Renaissance—which was the needed answer to the antiscience, pro-repression gloom of the Church-inspired Dark Ages—the world again awoke.

Take a moment—can you even imagine circuit-bending in a Dark Ages world? If caught, you literally would have been forced from your home and burned alive or flung from the church tower along with all the town's cats and wart-doomed elderly. But during the Renaissance, science and art were freed from hundreds of years of theological domination and brought out into the light of a recovering society.

Accordingly, the theretofore unmatched support for the arts and sciences resulted in the immediate scientific discoveries the Renaissance is noted for. This period of unbridled inspiration also nurtured new thoughts of social freedom. Enlightenment philosophers envisioned a world without the repression of art and science. They imagined a society free of tyrannical, self-important leaders.

One rather difficult-to-ignore result of this great philosophical vision was the eventual realization of the United States of America via its battered colonists breaking from England, starving in Jamestown, but, against all odds, surviving. Not to say that then or today we've shaken the grip of tyranny—we have not. But for the sake of our arts, sciences, and society, we've tried to break the grip, and we must continue to try. Circuit-bending, and indeed most of the arts as we know them, cannot survive otherwise.

To be sure, circuit-bending is a Renaissance art form, a revolutionary art form, and owes much to free thinkers of the past. Freedom of expression within the arts will flourish, the sciences will be free of nonscientists' misguidance, and a period of great discovery will ensue as long as we remember the lessons of the Dark Ages and the following Renaissance. Nothing is more important for the advancement of culture, the arts, and the wondrous wave of discoveries yet to come. You, as a bender, ride the crest of this wave.

So the power of discovery extends beyond article, beyond the actual item discovered. The power of discovery, on the broad front, is also the power of progressive, creative thinking. But free thinking is only a start. To extend the power of creative thinking, we must teach. And the power of teaching reigns supreme. Why? Because in teaching, in sharing ideas, we all move forward together. Sharing ideas, spreading the power of free, new, challenging, original thought, works. For you, me, and everyone. You're holding the proof in your hands.

You might find yourself in a similar spot someday: Do I share this idea? Do I teach? Or will this be my secret? As an inventor, I'm familiar with that battle. I decided to give away all the bending secrets because I believe in art more than I do corporate restraint, the needed, yet cold, underpinning of licensed intellectual property—the paranoia of plutocracy.

I find that I can't ignore the power of art to push discovery, the power of art to inspire new thought, and the power of art to teach (the very reasons that others repress the arts). I feel exceptionally fortunate to have been able to play this role of discoverer and teacher. Still, the real measure of fortune is in the reflection of personal deeds. To this end I've tried my best to provide a good handbook for your journey along what must be a somewhat rocky road. I sincerely hope you've enjoyed our trip, bumpy as it's been.

What next? Most readers of this book are musicians as well as designers. As such, your next step, after completing a nice set of instruments, will be composing music. Here is where the greatest satisfaction awaits. It is, after all, music that brings critical attention to instruments. If one listens (and designs) beyond the easy-grab distortion bends, instead looking to circuit-bending's delicate nuances, detailed voices, new timbres and fine musical languages, one will find room for insightful, sensitive composition. The promise of fine new music is as real as the instruments you've already made.

Let me introduce one last term: *theatric listening.* If you close your eyes when you listen to sound, any sound, and let your mind's eye illustrate the sound, this visualization may help you compose with the sound. This is theatric listening. It may take a little practice. But if you lose yourself within a sound or a music, your closed-eye field of vision will not remain blank. Something will appear. For me, this something is instructive and, if you'll bear with me here, it tells me what it needs. I imagine the accompanying sounds just as I might imagine the harmony to a singer's lyrics in a more traditional song. But that's just me—one of the ways I compose with experimental, unusual sound forms. You'll soon develop your own personal style of playing and composing with the bent instruments you've designed.

Wanting to learn more about electronics is a natural outgrowth of circuit-bending. I highly encourage further research and study. Bending and true-theory electronics are actually good friends. No matter how bent your circuits are, when you use the generic line-output scheme, you're already incorporating true-theory technique. The two approaches to design, although outrageously dissimilar, are yet complementary. And as a bender, you have a unique advantage to entering "real" electronics: not only experience, but also honest inspiration.

My closing idea for you is that bending, at its best, is a multi-use attitude. It's a thought process, not just an electronic technique; it's the Swiss Army knife of contemplation. The "what if?" postulation is just as powerful as the "here's how" surety. After you give yourself over to "what if," you naturally unlock everything possible. You leap the hurdles of conventional wisdom and see things anew. Suddenly, you're not only thinking outside the box. The box is gone. And the world, in a very personal way, becomes uniquely yours. Practice "what if?" and you'll see.

Now I'll stop the storytelling, the wild speculations, the unforgivable overstatements, and even the dreaming. I'll tell you now, quite down to earth, what I tell my apprenticeship students when we part: You were already a strong artist before we met. You were already capable. Your dedication to difficult art proves that, all by itself. The rest can only fall into place.

My guidance will amount to little in the end—just a signpost pointing toward your personal unknown. The circuits themselves will become your teachers, just as they became mine. Still, I hope I've helped you on your current journey, and that in your future unknown perhaps a foothold or two from this book will assist.

Thanks for letting me pull your leg, share some adventures, and be your guide. May you carry peace in your mind, love in your heart, and extra batteries in your pockets until the end of time.

One Dozen of the Best Generic Bends

Over the nearly forty years I've been bending, a number of very powerful "generic" bends have surfaced—bends that can be applied to many different circuits and with great results. Collected together in this appendix are one dozen of the best of these bends. All are extremely simple to understand and outrageously easy to implement. If you study these bends a little, until you grasp what each is up to, they'll spring into your memory as you explore circuits, cracking your new instrument possibilities wide open.

Many of the principles in this appendix have been touched upon in the main text of this book, and it is there you'll find more information. Here, however, you'll see things packaged "to go," for a quick reference and a jump-start into action. These are the "secrets of bending" that people talk about. But they're really just commonsense applications of the anti-theory and true theory techniques that hold bending together, finally brought out into the light, in their most basic form. Very powerful stuff. Have fun!

1. Line Output

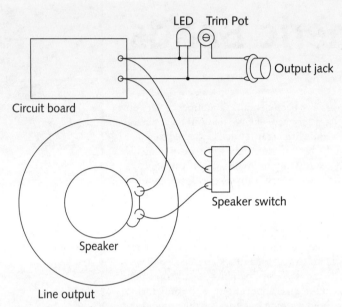

FIGURE A-1: Use this generic wiring scheme to derive a line output from your built-in speaker. The LED might not light (try various LEDs), and you might not need the trimmer. The speaker switch allows you to turn the speaker off, good for studio work. Be aware that switching the speaker off may increase the power of the line output. A nice variation on the output wiring is to use a full-sized potentiometer instead of a trim pot. Mount the full-sized pot right on the instrument and you have a handy output level control. Keep track of your polarity throughout the wiring. Use any style output jack desired.

2. "1 to X" Bend

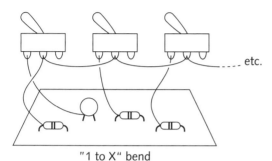

"1 to X" bend

Figure A-2: Many of the circuits in this book look to the "1 to X" bend for their basic bent functions. The "X" number is the direct result of how many good bends the traveling end of your bending probe uncovers as it searches the circuit. Accordingly, the "1" in the "1 to X" equation represents the circuit point that the stationary end of the probe rested on while the traveling end of the probe made its tour. Any number of points may be discovered by the traveling end; the final number of switches needed to implement the "X" side of things often exceeds the area available to mount them. Also see the patch bay drawing in this appendix.

3. "X to X" Bend

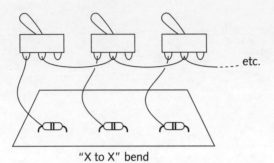

"X to X" bend

FIGURE A-3: As seen in the Harmonic Window project, this variant of the "1 to X" bend is useful for connecting many points together, but this time without there being a set single point common to all switches (point "1" in the "1 to X" bend). Within the "X to X" bend, the "1" from the "1 to X" bend simply becomes a variable. If all switches are turned off, no connection is made to the circuit. But if any of the switches are turned on, that switch then becomes the "1" in the traditional "1 to X" bend: any further turning on of switches will connect their circuit points to the point of the switch (the variable "1" switch) turned on prior. This is a cool elaboration on the traditional "1 to X" bend, often allowing a wider range of bent responses.

4. Bending Through a Potentiometer

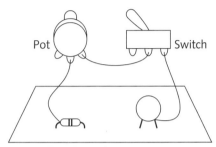

Pot   Switch

Bending through a Potentiometer

FIGURE A-4: Any time you discover a nice bend with your bending probe, remember that that bend might be variable if you send that connection through a potentiometer. There is no set potentiometer value to try here. Large value pots (above 1M) may be too coarse in their resistance change for a bend that needs only a slight resistance change to shine. On the other hand, a small value pot (5K or less) might not present enough change in resistance to alter the bend to the degree needed. I usually begin with a 1M potentiometer. If the bent response cuts off near the bottom (low resistance) of the pot's range, I know I need a pot with a smaller resistance value (like a 10K). If I turn the pot to full resistance and the bent effect is still changing nicely, I try a pot with a larger value (like a 10M). The switch in the diagram allows the potentiometer bend to be turned on or off, as usual.

5. Two-Way Choice

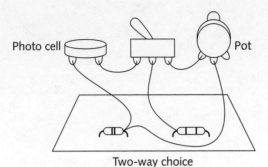

Photo cell   Pot

Two-way choice

FIGURE A-5: You might remember that I discussed
the exchangeability of toggle switches back in the
introduction to the projects chapters. So, what *do* you
do with a three-pole toggle switch if you want to use
all three poles instead of using only two to duplicate
the action of a two-pole (SPST) switch? As seen in the
diagram, a three-pole toggle switch (SPDT) is designed
to allow a signal to be directed down one of two
available paths. The signal enters the center terminal
of the three and exits through one of the two remaining,
depending upon which way the handle of the switch is
thrown. This allows you to select one of two components
to run your bend through. In the diagram you'll see that
the bend can now be switched between two variable
resistors: a potentiometer or a photo cell. Any two
components can be used.

6. Pitch Dial and Body-Contacts

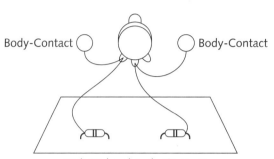

Body-Contact ⬭   ⬭ Body-Contact

Pitch Dial and Body-Contacts

FIGURE A-6: A very common way to set the "speed" of a circuit is by means of a well-placed resistor. The resistor governs the flow of electricity from one point to another, thereby establishing overall frequency response, or pitch. But are there other points on the board that will affect overall circuit speed, or pitch, if connected to each other? As you've discovered via body-contacting the circuit, yes, there are. As you've learned, wiring a potentiometer between such body-contact points often provides such a pitch control. But does this mean the body-contact response is overthrown by the introduction of a potentiometer? Not at all. As shown in the diagram, wiring body-contacts to the lugs of the potentiometer allows you the advantage of both. You can set overall pitch via the pot, and you can still touch the body-contacts to induce real-time vibrato and pitch-bending.

7. Chance Trigger

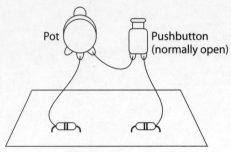

Pot   Pushbutton
       (normally open)

Chance Trigger

FIGURE A-7: Not all bends need to be kept connected to produce an interesting bent response. As you explore your circuits you'll eventually find two points that when briefly connected will initiate an on-going disruption of circuit activity (as seen in the SA-2 Aleatron in the projects chapters). The brief connection needed to create this response must be of the correct power to best initiate the unusual audio activity. Too much juice flowing through the connection equals a crash; too little juice, and nothing happens at all. Notice the potentiometer in the drawing. This pot regulates the amount of current flowing through the bend. Once it has been set correctly, a quick tap of the partner push button switch (a normally open mini switch) will apply the needed mini-zap to set the bent circuit response off.

8. Trimmed Potentiometer

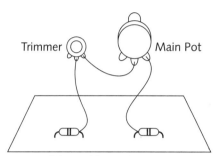

Trimmed Potentiometer

FIGURE A-8: Don't be surprised if you find a cool spot for a potentiometer to either adjust pitch or to vary some other interesting effect, only to discover that when turned all the way down (presenting no resistance to the circuit) the circuit quickly crashes. What's needed is to ensure that there's still a little resistance left in the bend when the pot's turned all the way down—to zero resistance. As shown in the diagram, the answer is simple: Once the trimmer is installed in line with the main potentiometer, all that's left is to set the trimmer to the resistance needed to keep the circuit from "seeing" zero resistance and crashing when the pot is turned all the way down. Turn the trimmer up all the way; turn the main pot down. With the bend running, turn the trimmer down until just pre-crash. This sets a safe operational minimum resistance for the bend.

9. Audition/Hold Array

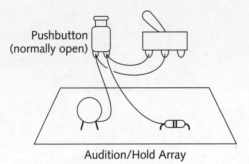

Pushbutton
(normally open)

Audition/Hold Array

FIGURE A-9: If you've built an Incantor, you're familiar
with how the looping controls work. You use the
pushbutton to audition loops, pressing it over and over
again to hear the loops available. Once you find a loop
you want to work with, you throw the companion toggle
switch to lock the loop in place. But beyond this, in
simply playing notes or sequences of bent sounds, it's
often nice to be able to choose between momentary or
continuous operation of such a sound (as seen in the
curved bottom row of switches on the Harmonic Window
instrument shown on the cover of this book). Wired exactly
the same as the looping matrix in an Incantor, this paired
pushbutton/toggle switch combination increases the
flexibility of sound production and performance.

10. Resistor Replacement

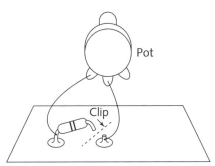

Resistor Replacement

FIGURE A-10: Resistors within a circuit, while being a very basic electronic component, are responsible for keeping much complex circuit activity within balance. Replacing a resistor with one of a different value can shift circuit activity from norm to alien in a moment. So imagine what replacing a resistor with a potentiometer or photo cell can do! Replacing a resistor need not be a hassle. In fact, complete physical removal is usually unnecessary to electronically remove the resistor from the circuit. In the diagram you see a resistor with one of its leads snipped just above the board. Even though still attached to the board, the resistor is now "invisible" to the circuit because electricity can no longer travel through it. Soldering your replacement component between the still-intact lead and the circuit board stub of the original resistor's clipped lead effectively bypasses the resistor, and with a minimum of hassle. If the substitution doesn't work out, re-soldering the single clipped lead of the target resistor will bring the circuit back to normal for further bending.

11. Reset Button

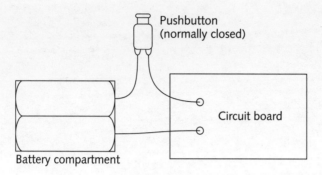

Reset Button

FIGURE A-11: Many bent circuits crash under the new electronic pressures of circuit-bending. Simply turning the main power switch off and back on might remedy such a crash. However, there are times when interrupting the current flowing from the actual power supply is the only way to reset a deeply crashed circuit. The solution is simple if the battery compartment is connected to the circuit by means of two wires, as is usually the case. Instead of removing a battery and again replacing it in order to interrupt the power supply and therefore reset a deep crash, simply breaking the circuit between the battery compartment and the circuit board is the answer. In the illustration you'll see depicted a normally closed pushbutton switch in the middle of one of the battery compartment wires. Such a switch on either of the two wires will interrupt the power supply current when pressed, instantly resetting the crash.

12. Patch Bay

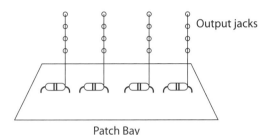

Patch Bay

FIGURE A-12: Instead of the X to 1 or X to X bends seen in this appendix and implemented in various of the projects in this book, a simple patch bay will also allow you the same, and even greater, connection capabilities. In the drawing only four circuit points are brought forward into patch points, and these points are each expanded to only four outputs within the patch bay. Naturally, any number of circuit points can be extended into a patch bay, and each point can open up into as many outputs as desired. If you're not pressured for space you can go with the traditional "banana" jacks and plugs for designing your bay. If space is limited (as it often is within circuit-bending) the alternative is the "pin" jack, a smaller and a little less sturdy substitute. In either example, parts are available for easily making your own patch cords and bay, one of the simplest ways to implement multi-connectivity within bending.

Benders' Resources

In this appendix, you'll find a great list of resources for circuit-bending whether you're looking for discussion groups, electronic parts, or tools. If bending has tempted you to construct your own amplifier, mixer, or other such circuit, you'll also find here a list of electronic kit sellers. Many of the dealers will supply you with free catalogs of their merchandise if you ask.

This appendix contains a collection of my personal favorites as well as recommendations by people deeply into conventional electronic prototyping as well as circuit-bending. My thanks to Bob Bamont (www.bobbamont.com/elcats.html) and Bill Beaty (http://amasci.com/amateur/amform.html) for the use of their on-site resource pages. Both sites are brimming with additional valuable information and both are highly recommended!

My Web site has an extensive set of benders' resources, including all those that follow, hyperlinked and ready to click. You'll find that gateway here: www.anti-theory.com/links/.

Benders' discussion groups:

- http://groups.yahoo.com/group/benders/
- http://groups.yahoo.com/group/bendersanonymous/
- www.em411.com/list/forum
- www.machinenoise.com/cgi-bin/YaBB/YaBB.pl/

Great circuit-bending link pages:

- **From Burnkit2600:** http://burnkit2600.machinenoise.com/links.html
- **From Cementimental:** www.cementimental.com/links.html
- **From Anti-Theory:** www.anti-theory.com/links/

Radio stations supportive of bent music:

- **WFMU 90.1FM (New Jersey, USA):** www.wfmu.org/
- **Resonance 104.4FM (London, UK):** www.resonancefm.com

Body-contacts:

(Threaded spheres and half-spheres)

Liberty Brass
38-01 Queens Boulevard
Long Island City, NY 11101
www.libertybrass.com/

(Metal knobs)

- **Knobs-n-Pulls:** www.knobs-n-pulls.com/iron-drawer-pulls-wholesale-we.htm

- **Knob Depot:** http://knobdepot.com/

(Threaded studs and spikes)

IMOSH
P.O. Box 497
Dept. WB1
Broadway, NJ 08808
www.imosh.com/

- **Studs and Spikes:** www.studsandspikes.com/

General electronic components:

ACE Electronics
1810 Oakland Road, Suite C
San Jose, CA 95131
www.acecomponents.com/

American Design Components
6 Pearl Court
Fairview, NJ 07022
www.adc-ast.com/

Ace Wire and Cable Co.
3 Self Boulevard
Carteret, NJ 11377
www.acewireco.com/

American Electronic Resource
3505-A Cadillac Avenue
Costa Mesa, CA 92626
www.aeri.com/

All Electronics Corp.
P.O. Box 567
Van Nuys, CA 91408
www.allelectronics.com/

American Science Surplus
P.O. Box 1030
Skokie, IL 60076
www.sciplus.com

Allied Electronics, Inc.
7410 Pebble Drive
Fort Worth, TX 76118
www.alliedelec.com/

Anchor Electronics
2040 Walsh Avenue
Santa Clara, CA 95050
www.demoboard.com/anchorstore.htm

Alltronics
2300 Zanker Road
San Jose, CA 95131
www.alltronics.com/

Antique Electronic Supply
6221 S. Maple Avenue
Tempe, AZ 85283
www.tubesandmore.com/

Apex Jr.
3045 Orange Avenue
LaCrescenta, CA 91214
www.apexjr.com/

B.G. Micro
555 N. 5th Street, Suite #125
Garland, TX 75040
www.bgmicro.com

Brigar Electronics
7-9 Alice Street
Binghamton, NY 13904
http://members.aol.com/brigar2/brigar.html/

Bud Industries, Inc.
4605 East 355th Street
Willoughby, OH 44094
www.budind.com/

Cascade Surplus Electronics
8221 N. Denver Avenue
Portland, OR 97217
www.cascadesurplus.com

C & H Sales Company
2176 E. Colorado Boulevard
Pasadena, CA 91107
http://aaaim.com/

Circuit Specialists
P.O. Box 3047
Scottsdale, AZ 85271–3047
www.cir.com

Circuit Specialists Inc.
220 S Country Club Drive #2
Mesa, AZ 85210
www.web-tronics.com/

Comptronics, Inc.
6259 West 87th Street
Los Angeles, CA 90045
www.icparts.com/

Consumertronics
8400 Menaul Boulevard
NE Suite A199
Albuquerque, NM 87112
www.tsc-global.com/

Dalbani Electronics
4225 N.W. 72nd Avenue
Miami, FL 33166
www.dalbani.com/

Davilyn Corp.
1300 Pacific Avenue
Oxnard, CA 93033
www.davilyn.com/

DC Electronics (KITS)
2334 N. Scottsdale Road
Scottsdale, AZ 85257
www.dckits.com/

Debco Electronics, Inc.
4025 Edwards Rd.
Cincinnati, OH 45209
www.debcoelectronics.com/

Digi-Key Corp.
701 Brooks Avenue South
Thief River Falls, MN 56701
www.digikey.com/

Edmund Scientific
101 E. Gloucester Pike
Barrington, NJ 00807–1380
www.edsci.com

Electronic Goldmine
P.O. Box 5408
Scottsdale, AZ 85261
www.goldmine-elec.com/

Electronic Inventory Online
1243 W. 134th Street
Gardena, CA 90247
www.eio.com/

Electronic Surplus Co.
9012 Central Avenue S.E.
Albuquerque, NM 87123
www.surplus-electronics.com/

Electronic Surplus, Inc.
5363 Broadway Avenue
Cleveland, OH 44127
www.electronicsurplus.com

Electronix Express
365 Blair Road
Avenel, NJ 07001
www.elexp.com/

Excess Solutions
430 E. Brokaw Road
San Jose, CA 95112
www.excess-solutions.com/

Fair Radio Sales
1016 E. Eureka
Lima, OH 45802
www.fairradio.com/

Gateway Electronics Inc.
2220 Welsch Industrial Court
St. Louis, MO 63146
www.gatewayelex.com/

Government Liquidation
(locations in almost every state)
www.govliquidation.com/

Herbach & Rademan (H&R)
353 Crider Avenue
Moorestown, NJ 08057
www.herbach.com

Hoffman Industries
853 Dundee Avenue
Elgin, IL 60120
www.hoffind.com

H&R Company
16 Roland Avenue
Mt. Laurel, NJ 08054
www.herbach.com/

Hosfelt Electronics, Inc.
2700 Sunset Blvd.
Steubenville, OH 43952
www.hosfelt.com/

Jameco Electronics
1355 Shoreway Rd.
Belmont, CA 94002
www.jameco.com/

JDR Microdevices
1850 S. 10th Street
San Jose, CA 95112
www.jdr.com/

Ken's Electronics
2825 Lake Street
Kalamazoo, MI 49048–5807
www.kenselectronics.com

Leo Electronics, Inc.
P.O. Box 11307
Torrance, CA 90510–1307
www.excess.com/

MAI Prime Parts
(aka MILO Associates)
5736 N. Michigan Road
Indianapolis, IN 46228
www.websitea.com/mai/

Marlin P. Jones & Associates
P.O. Box 12685
Lake Park, FL 33403
www.mpja.com/

MCM Electronics
650 E. Congress Park Drive
Centerville, OH 45459
www.mcmelectronics.com/

Mendelson Electronics Company, Inc.
340 E. First Street
Dayton, OH 45402
www.meci.com/

MG Electronics
1177 Park Avenue
Suite 5, Box 124
Orange Park, FL 32073
www.mgte.com

Midwest Surplus and Electronics
112 12th Avenue, S.
Minneapolis MN 55415
https://secure.midwest-electronics.com/

Mouser Electronics, Inc.
1000 N. Main Street
Mansfield, TX 76063
www.mouser.com/

Newark Electronics
100 High Tower Boulevard
Pittsburgh, PA 15205
www.newark.com/

Ocean State Electronics
P.O. Box 1458
Westerly, RI 02891
www.oselectronics.com/

Parts Express
340 E. First Street
Dayton, OH 45042
www.partsexpress.com/

Pasternack Enterprises (connectors)
P.O. Box 16759
Irvine, CA 92623
www.pasternack.com/

Prime Electronic Components, Inc.
150 West Industry Court
Deer Park, NY 11729
www.primelec.com/

Radio Shack
One Tandy Center
Fort Worth, TX 76102
www.radioshack.com/

RA Enterprises
2260 De La Cruz Boulevard
Santa Clara, CA 95050
www.angelfire.com/free/proto.html

R&D Electronics
5363 Broadway Avenue
Cleveland, OH 44127
www.electronicsurplus.com

Reliance Merchandizing Co.
651 Winks Lane
Bensalem, PA 19020
www.relianceusa.com/

Saturn Surplus
3298 Rt. 147 N.
Millersburg, PA 17061
www.saturnsurplus.com/

Skycraft Parts and Surplus, Inc.
2245 W. Fairbanks Avenue
Winter Park, FL 32789
www.skycraftsurplus.com/

Small Parts, Inc.
13980 N.W. 58th Court
P.O. Box 4650
Miami Lakes, FL 33014–0650
www.smallparts.com/

Surplus Center
1015 West "O" Street
Lincoln, NE 68528
www.surpluscenter.com/

Surplus Sales
1502 Jones Street
Omaha, NE 68102
www.surplussales.com

Surplus Sales of Nebraska
1218 Nicholas Street
Omaha, NE 68102–4211
www.surplussales.com/index.html

The Surplus Shed
8408 Allentown Pike
Blandon, PA 19510
http://surplusshed.com/

Surplus Traders
P.O. Box 276
Alburg, VT 05440
www.surplustraders.net/a/

Timeline Inc.
2541 West 237 Street
Building E
Torrance, CA 90505
www.timeline-inc.com/lcd.html

Toronto Surplus & Scientific, Inc.
608 Gordon Baker Road
Toronto, Ontario, Canada M2H 3B4
www.torontosurplus.com

TSC
8400 Menaul Boulevard
NE Suite A199
Albuquerque, NM 87112
www.tsc-global.com/surplus-e/

Tucker Electronics
1717 Reserve Street
Garland, TX 75042–7621
www.tucker.com

Unicorn Electronics
1142 State Road 18
Aliquippa, PA 15001
www.unicornelectronics.com/

Weird Stuff Warehouse
384 W. Caribbean Drive
Sunnyvale, CA 94089
www.weirdstuff.com/

LEDs:

American Bright Optoelectronics Corp.
13815-C Magnolia Avenue
Chino, CA 91710
www.americanbrightled.com/

Bivar, Inc.
4 Thomas Street
Irvine, CA 92618
www.bivar.com/

Chicago Miniature Lamp, Inc.
147 Central Avenue
Hackensack, NJ 07601
www.chml.com/

Clairex Technologies, Inc.
1845 Summit Avenue
Plano, TX 75074
www.clairex.com/

Cree, Inc.
4600 Silicon Dr.
Durham, NC 27703
www.cree.com/

Dialight Corp.
1501 Route 34 South
Farmingdale, NJ 07727
www.dialight.com/

Gilway Technical Lamp
55 Commerce Way
Woburn, MA 01801
www.gilway.com/

Kingbright Corp.
225 Brea Canyon Road
City of Industry, CA 91789
www.us.kingbright.com

Wm. B. Allen Supply Company, Inc.
301 N. Rampart Street
New Orleans, LA 70112
www.wmballen.com/

World Wide Wire
94 Via San Marco
Rancho Mirage, CA 92270
www.worldwidewire.com/

LEDtronics, Inc.
23105 Kashiwa Court
Torrance, CA 90505
www.ledtronics.com/

Lumex, Inc.
290 E. Hellen Road
Palatine, IL 60067
www.lumex.com/

Marktech Optoelectronics
120 Broadway
Menands, NY 12204
www.marktechopto.com/

Opto Technology
160 E. Marquardt Drive
Wheeling, IL 60090
www.optotech.com/

Rohm Electronics USA, LLC
10145 Pacific Heights Boulevard
San Diego, CA 92121
www.rohmelectronics.com/

Stanley Electric Sales of America, Inc.
2660 Barranca Parkway
Irvine, CA 92606
www.stanley-electric.com/

SunLED Corp.
20793 E. Valley Boulevard #C
Walnut, CA 91789
www.sun-led.com/

UnitedPRO, Inc.
10010 Pioneer Boulevard
Santa Fe Springs, CA 90670
www.unitedpro.com/

Electronic project kits:

All Electronics Corp.
P.O. Box 567
Van Nuys, CA 91408
www.unitedpro.com/

Almost All Digital Electronics
1412 Elm Street S.E.
Auburn, WA 98092
www.aade.com/

ApogeeKits
P.O. Box 625
Frisco, TX 75034–0625
www.apogeekits.com/

Boondog Automation
c/o Paul Oh
7709 Beech Lane
Wyndmoor, PA 19038
www.boondog.com/

Cal West Supply, Inc.
31320 Via Colinas
Westlake Village, CA 91362
www.hallbar.com/

Carl's Electronics
P.O. Box 182
Sterling, MA 01564
www.electronickits.com/

DC Electronics
2334 N. Scottsdale Road
Scottsdale, AZ 85257
www.dckits.com/

Edlie Electronics
2700 Hempstead Turnpike
Levittown, NY 11756
www.edlieelectronics.com/

ElectroKits
www.electrokits.com/

Electronic Goldmine
P.O. Box 5408
Scottsdale, AZ 85261
www.goldmine-elec.com/

Electronic Rainbow, Inc.
6227 Coffman Road
Indianapolis, IN 46268
www.rainbowkits.com/

Graymark
P.O. Box 2015
Tustin, CA 92681
www.graymarkint.com/

Kits R Us
http://kitsrus.com/

LNS Technologies
Box 501
Vacaville, CA 95696 USA
www.techkits.com/

Lynx Motion, Inc.
104 Partridge Road
Pekin, IL 61554
www.lynxmotion.com/

Mondo-tronics, Inc.
4286 Redwood Highway
San Rafael, CA 94903
www.robotstore.com/

Montek Electronics
695 Markham Road
Unit #18 (Second Floor)
Toronto, Ontario, Canada MIH 2A5
www.montek.com/

Ocean State Electronics
P.O. Box 1458
Westerly, RI 02891
www.oselectronics.com/

PAiA Electronics, Inc.
3200 Teakwood Lane
Edmond, OK 73013
www.paia.com/

Ramsey Electronics, Inc.
793 Canning Parkway
Victor, NY 14564
www.ramseyelectronics.com/

Ten-Tec, Inc.
1185 Dolly Parton Parkway
Sevierville, TN 37862
www.tentec.com/

Zagros Software
P.O. Box 460342
St. Louis, MO 63147
https://www.zagrosrobotics.com/Index.asp

Theremin World
65 Greenwood Avenue
Midland Park, NJ 07432
www.thereminworld.com/shop_theremins.asp

Specialty paints:

- **Righter Group:** www.rightergroup.com/product_info/specialty_paints.asp

- **Krylon:** www.krylon.com/

- **Plastikote:** www.plastikote.com/

- **Sargent Art:** www.sargentart.com/

- **Rustoleum:** www.rustoleum.com/

- **Glo Paint Pro** (luminous): www.glowpaintpro.com/

- **Clear Neon** (invisible fluorescent): www.clearneon.com/

Paint booths:

- **Alex Kung's DIY Booth:** www.interlog.com/~ask/scale/tips/booth.htm

- **Donald Granger's DIY Booth:** http://donaldgranger.home.att.net/paint_booth.htm

Soldering stations (recommended Weller WLC 100)

- **Action-Electronics:** www.action-electronics.com/westations.htm

- **Consolidated Electronics:** www.ceitron.com/solder/station.html

- **JDR Microdevices:** www.jdr.com/interact/item.asp?itemno=WLC100

- **Ocean State Electronics:** www.oselectronics.com/ose_p60.htm

Electronics tools:

Contact East, Inc.
335 Willow Street
North Andover, MA 01845
www.contacteaStreetcom/

Jensen Tools, Inc.
7815 S. 46th Street
Phoenix, AZ 85044
www.jensentools.com/

HMC Electronics
33 Springdale Avenue
Canton, MA 02021
www.hmcelectronics.com/

Specialized Products Co.
1100 S. Kimball Avenue
Southlake, TX 76092
www.specialized.net/ecommerce/shop/

Techni-Tool, Inc.
1547 N. Trooper Rd.
Worcester, PA 19490
www.techni-tool.com/

Tecra Tools, Inc.
2452 S. Trenton Way
Denver, CO 80231
www.tecratools.com/

Time Motion Tools
12778 Brookprinter Place
Poway, CA 92064
www.timemotion.com/

Travers Tool Co., Inc.
128-15 26th Avenue
College Point, NY 11354
https://www.travers.com/index.asp

Victor Machinery Exchange, Inc.
251 Centre Street
New York, NY 10013
www.victornet.com/

W. S. Jenks & Sons
1933 Montana Avenue N.E.
Washington, DC 20002
www.wsjenks.com/

Wassco
12778 Brookprinter Place
Poway, CA 92064
www.wassco.com/

Weird electronics (some kits, too):

- **Bull Electrical:** www.bull-electrical.com/

- **Cliveyland:** http://bigclive.com/

- **Don Klipstein's Odd Bulb page:** http://members.misty.com/don/oddbulb.html

- **Electric Stuff:** http://electricstuff.co.uk/

- **The Electrotherapy Museum:** www.electrotherapymuseum.com/

- **ePanarama Electronic Resources Links:** www.epanorama.net/links/circuitsites.html

- **Free Energy News:** www.freeenergynews.com/

- **Future Horizons:** www.futurehorizons.net/

- **Information Unlimited:** www.amazing1.com/

- **Lone Star Consulting:** www.lonestartek.net/weird.htm

- **Science Hobbyist:** www.eskimo.com/~billb/

- **Secret Technology Labs:** www.espionage-store.com/secret/index.html

Experimental instrument sites:

- **EMI:** www.windworld.com/

- **The Oddmusic Site:** www.oddmusic.com

- **My Web site** with How-to's, Instrument Galleries, and more: www.anti-theory.com/;
  E-mail: ghazala@anti-theory.com

- **Dan Stowell's Oddmusic Links Archive:** www.mcld.co.uk/search/oddmusic

Common Electronic Components and Their Schematic Symbols

appendix C

First of all, no, you don't need to know anything in this appendix in order to explore circuit-bending. Still, you'll be running into schematic diagrams as you move deeper into electronics, and you might even want to chart your own bends using the traditional symbols of the field. And these symbols exist for a very good reason.

Circuitry can get complicated. To answer the demands of documenting (and reading) the complex wiring of circuits, the schematic diagram system was developed. The goal was to simplify not only component depiction, but wiring depiction as well. If the system really works, the drawings in this appendix should aid in your understanding of the most common components appearing on the circuit boards you'll be bending, as well as how they're traditionally depicted within circuitry. But in no way is this a complete list of all the components available, or their various cryptic renderings. (That set on the Roswell saucer still has everyone stumped.)

So, I've limited my depictions to the best-known and most common standards of schematic drawings. These are accompanied by my own observations of these drawings in actuality (where discrepancies abound). So let this appendix serve only as an introduction to the cool codes of circuitry. And feel free to design your own symbols as you incorporate or manufacture new components to suit your project's needs, as you'll see I've had to do with one of the most common components in bending: the simple metal body-contact.

In the example of the early all–body-contact–instrument I built back in the 1960s, you might recall that a group of semi-clad people could all be configured as body-contacts if everyone held hands and touched the instrument—while they touched each other here and there. A schematic symbol for that body-contact I have not yet figured out.

Resistors

Resistors are familiar to anyone who's ever looked at circuitry and may be the most common component in circuit design. Circuit-board–mounted resistors can range in size from minute, as in SMD (surface mount device) configuration, to humongous (way bigger than a breadbox, whatever that is). Resistance rating is coded in color bands painted on the resistor itself (pick up a resistor color wheel at your electronics store for easy reference). In diagrams, resistors are always represented by a zigzag line. The value of the resistor (its level of resistance) might be noted near its picture. If not, a coded label will appear next to the resistor's symbol (e.g., R1— the letter "R" is used to label resistors in schematic diagrams). This code can then be looked up in the schematic diagram's master parts list where the actual value of the resistor will be stated.

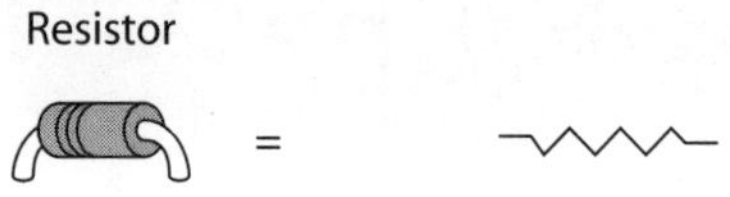

Resistor

Figure C-1

Potentiometers

Potentiometers are variable resistors whose values are stated in terms of the greatest resistance the "pot" is capable of presenting to the circuit (with the shaft turned all the way). Similar to that of a fixed resistor, the schematic symbol of a pot also contains a zigzag line (representing the body of actual resistance material within the potentiometer). But added to the zigzag is an arrow representing the movable "wiper" within the pot—the metal arm that "taps" electricity off the varying lengths of resistance material the current is running through as you turn the wiper's shaft. If only two lugs on the pot are used in the wiring (middle and one of the outside lugs) you'll see in the schematic diagram that one end of the zigzag is left unconnected to the circuit. The letter "R" is used to label pots in schematic diagrams, just as with resistors. After all, pots are resistors too.

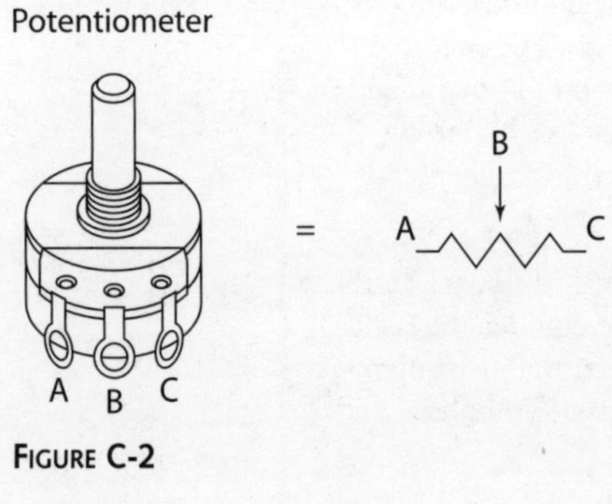

Potentiometer

Figure C-2

Photocells

Photocells are also variable resistors, and their schematic symbol also contains the familiar zigzag line associated with resistors. Along with an arrow similar to the arrow representing the wiper of a potentiometer (the pot's variable component), extra arrows now indicate a photocell's variable component: light. The arrows represent sunlight falling upon the cell, making the resistance drop. In some cases a value will be indicated next to the cell's symbol (or even a part number), but often you'll see a schematic symbol for a photocell with no title at all. A "generic" photocell is used, or, better yet, you'll substitute a number of cells and see which one works best (photocells vary in sensitivity, resistance range, and response time). You'll often find "Photocell" spelled out next to its symbol in schematic diagrams. I've also seen "R" (for resistor) and "Cell" and "CdS" (for cadmium sulfide, the material of common photocells) courting the symbol as well.

Photocell

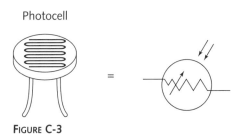

FIGURE C-3

Capacitors

Capacitors come in a variety of packages. The two most familiar are disc (looking like a disc) and electrolytic (looking like a cylinder with one end crimped). Like resistors, "caps" range in size from tiny to huge. Some capacitors are coded with color bands, but most have their value marked clearly on the body in plain ol' Greek (and English, thankfully), as in "47 μF," a measure of their capacitance. While disc caps are nonpolarized and can be soldered into a circuit any old way, electrolytic caps must be wired into the circuit according to their polarity. Polarity will be indicated on the component as well as in the schematic diagram (a "+" sign will appear on the positive side of the capacitor's schematic symbol; a negative indicator often appears on the capacitor). The letter "C" is used to label capacitors in schematic diagrams.

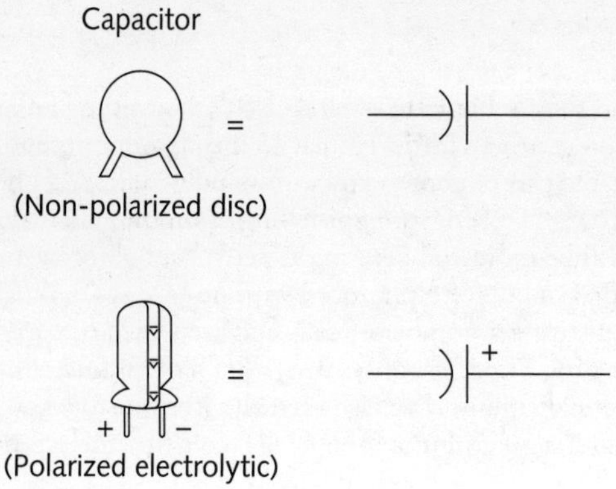

Capacitor

(Non-polarized disc)

(Polarized electrolytic)

FIGURE C-4

Crystals

Crystals set frequency in circuitry (often used to tune a master pitch or to clock general circuit activity). Their schematic symbol is close to that of the capacitor (each sets up pulses at assigned frequencies, if you're looking for a tie-in to support the similar graphic representations). There are two main types of crystal packages common to circuitry: the ceramic, looking like it could be just about anything (capacitors, resistors, and other components also come housed in such indistinct packages); and "canned" crystals, looking like small metal cans, rounded off a bit at the corners. Crystals are almost always given away by their rating, printed plainly on the side in MHz (megahertz). The letter "X" or the letters "XTL" might be present to identify a crystal, but you might find a code for the crystal (often including its MHz) as the only label present.

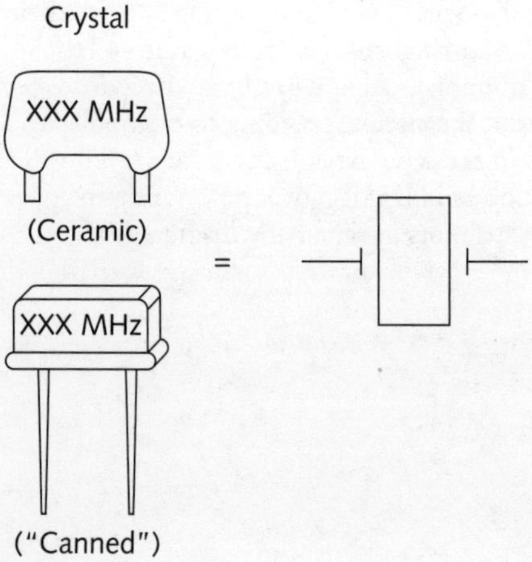

Crystal

(Ceramic)

("Canned")

FIGURE C-5

Transistors

Transistors are most likely the only three-legged component you'll encounter on the board. They're nothing more than semiconductor-based switches or amplifiers in many common applications. But their similarity stops there, because transistor package styles are many. Most common in battery-powered audio circuits are transistors housed within small gray-black cylinders, cylinders with one edge flattened. Depicted is the common symbol, but you'll encounter variations on this circle-and-three-lead picture as you see representations of photo transistors (with little arrows pointing into the symbol, as with the photocell) and other transistor styles. Circuit-benders rarely incorporate new transistors in the bent wiring scheme, but you may have the need to solder to a transistor lead now and then (remember to use a heat sink). The letter "Q" is used to designate transistors in schematic diagrams.

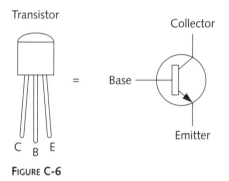

FIGURE C-6

Integrated Circuits

Integrated circuits (ICs) are always pin-numbered counting counterclockwise from the uppermost pin of the case's left-hand row, with the IC situated vertically, top up (the top of an IC is indented or notched to aid in pin identification). However, for the sake of clarity (fewer crossing wires on the drawing), schematic diagrams of ICs often rearrange pins emerging from the IC. So don't expect the counterclockwise numbering to hold true in schematics. The good news is that IC pins on a schematic drawing are all numbered, and the diagram will show where they're all meant to go. In some digital circuits, "unused" IC pins must still be connected, or "tied," to ground. You'll remember that such pins are all meant to be connected to the negative (-) side of the power supply. The letters "IC" identify integrated circuits in schematic diagrams, and they are followed either by the actual IC designation (as in IC LM3909, and you'd better hope not—it's extinct) or by a number referring you to the master parts list.

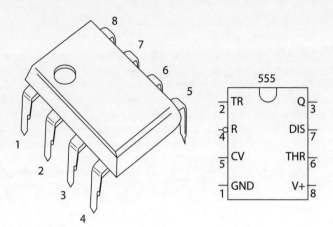

FIGURE C-7

Diodes

Diodes are one-way gates in electronics: They'll allow current to pass in one end and out the other, but not in reverse. In other words, diodes are polarized. On the component itself you'll find a printed band toward one end of its body, encircling the diode's cylinder. This band corresponds with the line at the tip of the arrowhead in the diode's schematic symbol.

The arrow points in the direction the diode will allow electricity to pass through it, in the unmarked end and out the end marked by the band. So a diode held against the schematic symbol of itself will be correctly oriented when the band on the diode and the line at the diagram's arrowhead correspond (in the drawing, band and line would be at the diode's right-hand side). The letter "D" is used to identify a diode in a schematic diagram, usually followed by a number allowing the reader to refer to the master parts list for further identification.

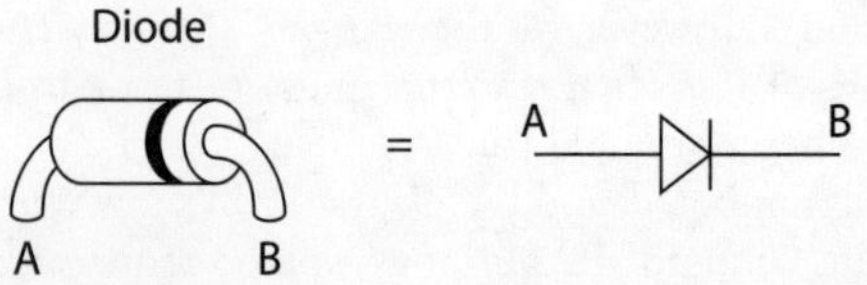

Diode

A   B   =   A ———▷|—— B

FIGURE C-8

LEDs

LEDs (light emitting diodes) are represented in schematic diagram in much the same way as regular diodes. But in addition to the standard schematic drawing of a simple, non-illuminating diode, arrows are now added to indicate the emitting of light. Because LEDs are polarized components, reversing their leads in a circuit will keep them from lighting. So, just the same as with regular diodes, LEDs must be oriented according to their polarity, and their polarity is indicated within schematic diagrams. Many LEDs are available, and their voltage requirements differ significantly. Within a similar voltage range they're interchangeable. But step outside that range and you risk burnout. The letters "LED" identify an LED within a schematic diagram and lead you to its specific part number or description in the master parts list.

LED

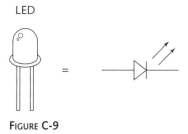

FIGURE C-9

Tungsten Light Bulbs

Tungsten (incandescent filament) light bulbs are appearing in fewer and fewer circuits, rapidly being replaced by the more electronically efficient LED. I'm very fond of tiny tungsten lamps (called grain-of-wheat bulbs) and incorporate them into instrument designs. While LEDs are available in "white" (there's a violet edge to the white) and can be painted with a transparent colored gloss, coloring in the same way a tiny, brilliant white tungsten illumination source gives a much more striking point of colored light. In schematic diagrams, incandescent lamps, unfortunately, look like unhappy circles. The frown, of course, represents the bulb's filament, and in that sense the symbol is rather sleek. If the rating of the bulb (voltage /wattage) does not appear next to the symbol but instead you find (as usual) a part code (like L1, for Lamp 1), you're being referred to the diagram's master parts list, which will dispel this (and other such) vagueness.

Tungsten bulb

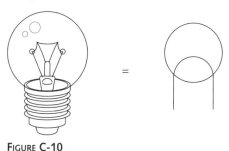

FIGURE C-10

Neon Lights

Neon lights are fun to experiment with and are used in electronic projects for indicator lights as well as in some equipment as frequency strobes (like in older pitch tuners—"Strobo-Tuners"— and as strobe lamps to help tune the speed of spinning record platters). The phosphor coating on the inside of a neon lamp's glass envelope (along with varying gases inside) can turn the usual orange glow into an assortment of other colors, green being the most common alternate hue. No matter what the color, the schematic symbol for neon lamps is always the same, as seen in the drawing. The letters "NE" often label a neon bulb in a schematic diagram, though more specific information (the code for the bulb type) may appear alongside the bulb as well. If not, such information will usually appear in the master parts list that accompanies all good schematics.

Neon bulb

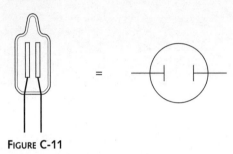

Figure C-11

Speakers

Speakers in schematic symbol are shown in side-view, although the artwork is anything but standardized. Usually speakers consist of a small rectangle (the magnet) and a funnel-shaped trapezoid (the cone) whose dimensions vary with the artist's whim. Often the symbol is lacking all identifiers other than an ohm rating (commonly 8Ω, but lesser and greater ohmage will be encountered). At other times "SPK" or "SPEAKER" will be used as a label. If a speaker diameter is of importance, this information will more than likely be included in the master parts list. Speakers are polarized components (although, unlike batteries and LEDs, they'll still work if wired in reverse, but they'll be "out of phase" with other speakers in the system wired correctly). Examine the speaker's two soldering terminals and you'll probably see polarity marks. Follow this information and wire the speaker as the schematic indicates.

Speaker

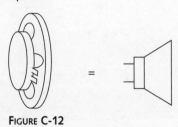

Figure C-12

Microphones

Microphones come in two basic styles, schematic-wise. These two types are passive (nonpowered) and powered (needing a voltage input to operate). Figure C-13 shows the schematic symbol of a passive microphone. If the microphone requires a separate power supply to operate, a third lead will be included in its schematic representation, and this lead will usually be labeled with the power required to operate the microphone. (The power supplied to this microphone may be drawn from a separate section of the circuit, or a separate power supply might be indicated as the microphone's power source.) The letters "MIC" usually label a microphone within a schematic diagram. Any specific information further describing the microphone's properties or requirements will be found in the schematic diagram's master parts list.

Microphone

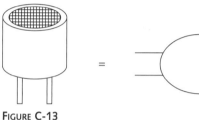

FIGURE C-13

Body-Contacts

I've seen very few schematic symbols for body-contacts over the years—none that I accurately remember. In my own drawings I use a simple circle with connecting wire included. Next to this circle I include the letters "BC." The most standard body-contact in circuit-bending is the "tapped ball," a product available in the lamp departments of larger hardware stores. There are, however, endless items available for use as body-contacts. If you create a schematic that includes a body-contact, I suggest you follow the usual abbreviations seen in schematic diagrams (like "BC1"), the final number referring you to the master parts list where you can specify a tapped ball of a certain size, a drawer knob, a threaded spike, or whatever you're using.

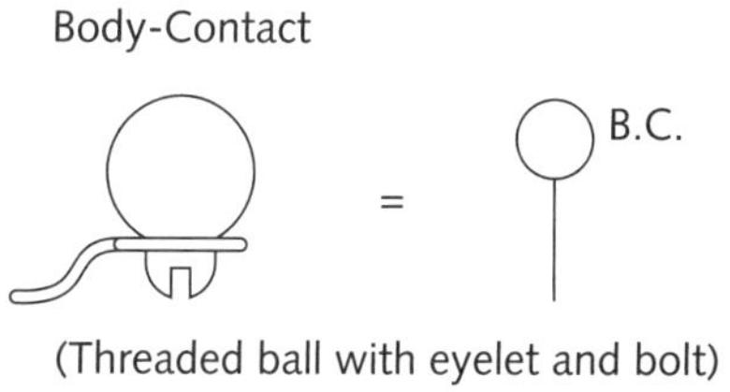

Body-Contact

B.C.

(Threaded ball with eyelet and bolt)

FIGURE C-14

Toggle Switches

Shown is the schematic symbol for the toggle switches used for building the instruments in this book's projects pages. This is the simple "SPDT" (single pole double throw) switch. You'll see many variations on this switch symbol as you read various schematic diagrams. The arrows will represent the moving segment of the switch. The dots or contacts to which the arrows point represent the internal contacts that the switch opens or closes as it operates. The letters "SW" label switches in schematic diagrams and refer you to the master parts list, where specific descriptions of switch type will be noted.

SPDT Toggle Switch

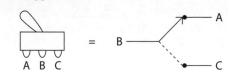

FIGURE C-15

Normally Open Pushbutton Switch

A normally open pushbutton switch is shown in schematic diagrams as a very simple (and sensible) side-view of its mechanical operation. Indicated are the plunger, the two inner contacts bridged by the plunger's conductor, and the conductor itself. The symbol always shows the pushbutton at rest (with the two inner contacts normally open, or not connected to each other). It takes but a little imagination to see the switch operating: If the plunger is depressed, the conductor will be lowered onto the two inner contacts, and the circuit will be closed. As with all switches, the letters "SW" will indicate their presence within a schematic diagram, followed by a number leading you to a detailed description in the master parts list.

Normally Open Pushbutton Switch

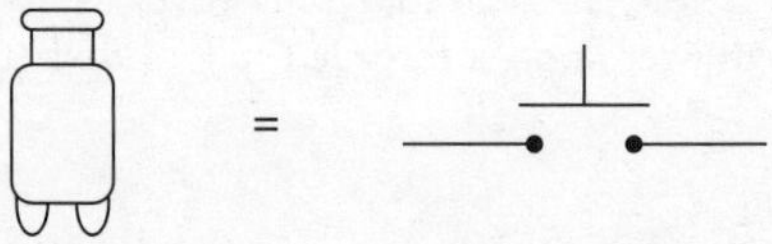

FIGURE C-16

Normally Closed Pushbutton

A normally closed pushbutton switch is shown in schematic diagrams as a very simple (and again, very sensible) side-view of its mechanical operation. Indicated are the plunger, the two inner contacts bridged by the plunger's conductor, and the conductor itself. As in the example of the normally open pushbutton switch, the symbol always shows the pushbutton at rest (with the two inner contacts normally closed, or connected to each other). It's easy to imagine the switch operating: If the plunger is depressed, the conductor will be lowered away from the two inner contacts, and the circuit will be broken. As with all switches, the letters "SW" will indicate their presence within a schematic diagram, followed by a number leading you to a detailed description in the master parts list.

Normally Closed Pushbutton Switch

FIGURE C-17

Rotary Switches

Rotary switches are depicted showing, primarily, the "common" terminal (the contact common to all switch positions) and the variable number of outer terminals that the signal current applied to the common terminal will flow through when the switch's dial is turned to that outer terminal's position. Just as in the example of toggle switches, rotary switches are designed in various configurations. You'll find from few to many contacts on the switch diagram, depending upon the switch's design. (Some rotary switches use stacked platforms of contacts to achieve dozens of contact changes with one click of the dial.) Again, the letters "SW" label switches in schematic diagrams and refer you to the master parts list.

Multi-Position Rotary Switch

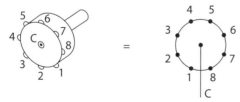

FIGURE C-18

One-quarter-inch Audio Jacks

Schematic diagrams of one-quarter-inch "guitar" audio jacks are, like pushbutton switches, side-views of the mechanical working parts. You'll see long extensions with a kink at the end; these represent the springy metal contacts that touch the tip section(s) of the plug when inserted into the jack. The common (or negative) side of the jack is the larger metal section—the round part that mounts to the panel of the instrument it's installed on. Taking a careful look at the actual jack will reveal which soldering lug goes to which part of the jack. The smaller one-eighth-inch jack is represented in schematic diagrams the same way, being identical in operation but on a diminutive scale. The letter "J" (or the word "Jack") usually accompanies the schematic symbol, although at times you'll find jacks labeled as simply "input" or "output."

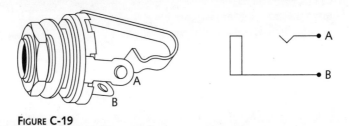

FIGURE C-19

RCA Jacks

As opposed to the guitar jack, whose schematic symbol is based upon a side-view of the mechanical characteristics of the jack, the RCA jack (or "phono" jack) is depicted as a very basic front-on view of only the visual characteristics. This results in a bull's-eye–like image, with the dot in the middle representing the central (positive, or "hot") terminal, and the outer ring portraying the remaining (negative, or ground) terminal. As long as the plug to be inserted into this jack is also wired correctly (positive to tip; ground to the remaining terminal), all's well. As with the guitar jack, the letter "J" (or the word "Jack") will appear alongside the symbol as a label. And, again similar to the larger guitar jack, you may find RCA jacks labeled in schematic diagrams only as "input" or "output."

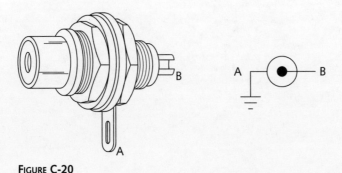

FIGURE C-20

Ground

Electrical "ground" is indicated by the dashed triangle. A convention to help simplify schematic diagrams, this symbol eliminates the clutter of all the lines otherwise needed to indicate where wires would attach to the common, or ground, side of the circuit (many points in circuits often connect to ground). But what *is* ground? While an "earth" ground is called for in some electronic applications (like grounding an antenna tower by actually running a wire from it to a spike driven into the earth), for us ground is simply the negative side of the circuit. Accordingly, all points on a circuit terminated in a ground symbol are to be connected to the negative (-) side of the power supply. In the example of the antenna, "earth ground" will be written next to the symbol, avoiding any confusion with regular grounding and keeping you from running a bunch of wires from your Speak & Spell out the window and into the garden.

GROUND (Connected to negative side of power)

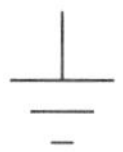

FIGURE **C-21**

Power Supplies

Power supply symbols may involve transformers, batteries, or simply two points whose power characteristics are declared (as in "6V DC"). Negative and positive sides of the power inputs will be declared in battery-powered circuits. (In alternating-current circuit diagrams you'll find either a simple "AC" declaration including a specific voltage [or range of voltages] or a more complete description entailing hot, common, and perhaps grounding information.) Remember that it's routine in schematic diagrams to omit much of the negative (common) side of the wiring that actually connects the negative side of the power supply to various points on the circuit (see "Ground," just previous in this appendix). These circuit points meant to be connected to the negative side of the power supply are marked in the schematic diagram with the ground symbol.

Power Inputs

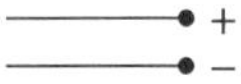

FIGURE **C-22**

Batteries

Batteries in schematic diagrams are usually shown as individual cells—but not always. You may run into a schematic of what looks like a single cell (as in Figure C-23), but with a label indicating six volts. Yes, this might be a single 6-volt battery. But it might also be a battery pack containing four 1.5-volt batteries. So you might need to read between the lines when looking at the suspicious battery schematic from time to time. The schematic diagram for a battery contains an indication of polarity. Usually this is in the form of two marks: a positive (+) sign at the longer parallel line of the symbol and a negative (-) sign at the opposite end. On occasion the negative sign is omitted. If an amperage rating is expressed, be sure to power the circuit with a supply that meets (or exceeds) this requirement. In simple terms, erring on the hi-amp side of this equation (4 "D" cells instead of 4 "AAA" cells to achieve your 6 volts) can't hurt.

Battery

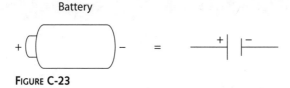

FIGURE C-23

Wiring

Whether or not wires touch each other within a circuit is pretty important to indicate—unless, of course, you enjoy the fragrance of really hot semiconductors. Several conventions are used to depict whether wires that cross are connected to each other. Quite straightforward: If they're connected you'll see a dot at their junction. Simple enough, but if they're *not* connected you'll see one of two things. In the good ol' days of electronics, when asbestos was used not only as insulation but also as chewing tobacco, unconnected wires were shown with one looping up over the other. Today you'll more often see their junction looking just the same as connected wires – but without the dot. Both styles will be encountered as you read schematics. While the looping style may no longer be in vogue, the advantage is in its indication that the author was actually thinking at the moment the line was drawn and didn't just forget to dot the intersection.

Wiring

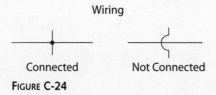

Connected   Not Connected

FIGURE C-24

Index

Continued

Continued

Continued

Continued

Continued

Continued

How to take it to the Extreme

If you enjoyed this book, there are many others like it for you. From *Podcasting* to *Hacking Firefox*, ExtremeTech books can fulfill your urge to hack, tweak, and modify, providing the tech tips and tricks readers need to get the most out of their hi-tech lives.